P9-AZX-989

single digits

single digits

In Praise of Small Numbers

MARC CHAMBERLAND

Princeton University Press

Princeton & Oxford

Published by Princeton University Press, 41 William Street, Princeton, New Jersey 08540
In the United Kingdom: Princeton University Press, 6 Oxford Street, Woodstock, Oxfordshire OX20 1TW
press.princeton.edu

Library of Congress Cataloging-in-Publication Data

Chamberland, Marc, 1964–
Single digits : in praise of small numbers / Marc Chamberland.
pages cm
Includes bibliographical references and index.
ISBN 978-0-691-16114-3 (hardcover : alk. paper) 1. Mathematical analysis. 2. Sequences (Mathematics) 3. Combinatorial analysis. 4. Mathematics–Miscellanea. I. Title.
QA300.C4412 2015
510—dc23 2014047680

British Library Cataloging-in-Publication Data is available

This book has been composed in Adobe Caslon Pro and Myriad Pro

Printed on acid-free paper. ∞

Typeset by S R Nova Pvt Ltd, Bangalore, India
Printed in the United States of America

1 3 5 7 9 10 8 6 4 2

contents

preface

> When you have mastered numbers, you will in fact no longer be reading numbers, any more than you read words when reading books. You will be reading meanings.
>
> —W. E. B. Du Bois

A colorful story concerns two mathematical wizards of early twentieth-century Britain, G. H. Hardy, the intellectually towering professor from Cambridge, and Srinivasa Ramanujan, the young genius from India. Their unlikely collaboration culminated in Hardy's invitation to bring Ramanujan to England to jointly pursue research. After a few years, Ramanujan fell ill—diagnosed with tuberculosis—and was confined to a sanatorium. Hardy recounts an unusual conversation during a visit to Ramanujan:

> I remember once going to see him when he was ill at Putney. I had ridden in taxi cab number 1729 and remarked that the number seemed to me rather a dull one, and that I hoped it was not an unfavorable omen. "No," he replied, "it is a very interesting number; it is the smallest number expressible as the sum of two cubes in two different ways" (Hardy, *Ramanujan*, p. 12).

Indeed, one can write $1729 = 1^3 + 12^3 = 9^3 + 10^3$. How does one come up with such observations? You have to study the numbers for a long time and make many connections. With talent, burning curiosity, and few distractions, Ramanujan became the master of numbers. J. E. Littlewood, Hardy's longtime collaborator, commented upon hearing the taxicab story, "Every positive integer is one of Ramanujan's personal friends."

This book is about single digits, specifically, the numbers 1 to 9. While it was tempting to also include the number zero, I decided to stick with the counting numbers. Each number has fascinating properties connected to many different areas of mathematics, including number theory, geometry, chaos, numerical analysis, mathematical physics, and much more. Some of the topics, such as the Pizza Theorem, require little mathematical background and are understandable by a curious 12-year-old; other sections require modest amounts of technical math, while a few sections, such as the section on E_8, allude to such sufficiently advanced material that it should not be read with small children present. Virtually every section is an independent vignette. For the cautious reader, the easier sections tend to lead off each chapter, and the earlier chapters are easier as a whole. Some of these topics are further explored in my YouTube channel, Tipping Point Math. There is assuredly something new and enlightening for everyone, the numbers 1 to 9 connecting to a multitude of mathematics in magical ways. I hope that you will come to call these numbers your friends.

I am grateful to my Grinnell colleagues Chris French, Joe Mileti, and Jen Paulhus and other mathematical colleagues Art Benjamin and Mike Mossinghoff for helpful advice. It's wonderful to have a network of bright, resourceful people who can offer thoughtful, timely answers when needed. Special thanks go to my editor Vickie Kearn at Princeton University Press for her support all along the way. Lastly, I want to thank my family, especially my wife Marion, for encouragement during the long writing process. All this help leaves me positively disposed for future projects.

single digits

1

The Number One

All for one and one for all.
— *The Three Musketeers* by Alexandre Dumas

Even if you are a minority of one, the truth is the truth.
— Mahatma Gandhi

The number one seems like such an innocuous value. What can you do with only one thing? But the simplicity of one can be couched in a positive light: *uniqueness*. We all know that being single has its virtues. In a mathematical world with so many apparent options, having exactly one possibility is a valued commodity. A search of mathematical research papers and books found that more than 2,700 had the word "unique" in their titles. Knowing that there is a unique solution to a problem can imply structure and strategies for solving. Some (but not all) of the sections in this chapter explore how uniqueness arises in diverse mathematical contexts. This brings new meaning to looking for "the one."

Sliced Origami

Origami traditionally requires that one start with a square piece of paper and attain the final form by only folding. When mathematicians

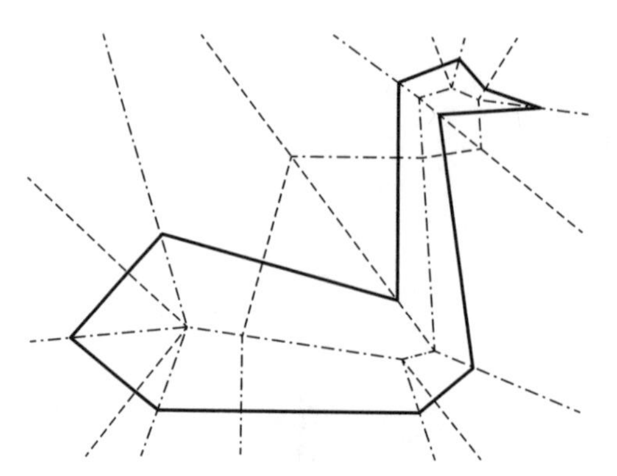

FIGURE 1.1: Origami fold pattern for a swan.

seriously entered the world of origami, they began to systematize constructions, including the use of computers to make precise fold patterns. Besides adding to the art, their contributions have also led to practical applications. For example, how can one transport a solar panel array into space? Mathematical origami has produced a design that compactly stores the whole assembly for transport. Once in space, the array is unfolded to its full size.

Children have long made beautiful paper snowflakes, but such constructions violate a fundamental origami tradition: no cutting, tearing, or gluing. But what if we were allowed a single cut? What new patterns could be attained? The surprising answer was found by Erik Demaine, a young Canadian professor at MIT, whose research intersects art, mathematics, and computer science. Demaine proved that *any* pattern whose boundary involves a finite number of straight line segments can be made by folding a paper appropriately and making a *single* cut! The possibilities include any polygon or multiple polygons. Figure 1.1 demonstrates the fold pattern needed to make a swan. After folding along each of the dashed and dashed-dot lines, the figure can be collapsed—with practice—so that a single cut along the bold line produces the swan.

Fibonacci Numbers and the Golden Ratio

The Fibonacci numbers are a sequence that has attracted attention from amateur investigators as well as seasoned mathematicians. As a reminder, the first two positive Fibonacci numbers are both 1, and each subsequent number in the sequence is formed by adding the previous two. This produces the sequence 1, 1, 2, 3, 5, 8, 13, 21, etc. Letting F_n denote the nth Fibonacci number, there is a tidy closed-form formula:

$$F_n = \frac{1}{\sqrt{5}}\left[\left(\frac{1+\sqrt{5}}{2}\right)^n - \left(\frac{1-\sqrt{5}}{2}\right)^n\right]$$

As n gets larger, the second term shrinks to zero, so F_n can be approximated as

$$F_n \approx \frac{1}{\sqrt{5}}\left(\frac{1+\sqrt{5}}{2}\right)^n$$

This shrinkage implies that the ratio of successive terms satisfies

$$\frac{F_{n+1}}{F_n} \approx \frac{1+\sqrt{5}}{2} \tag{1.1}$$

The constant on the right—usually denoted by the Greek letter ϕ—is called the *Golden Ratio*. Connections of this number to art, architecture, and biological growth have long been studied, but the connection between ϕ and the number 1 is not as well known. Two beautiful formulas make the link apparent. The first is

$$\phi = 1 + \cfrac{1}{1 + \cfrac{1}{1 + \cfrac{1}{1 + \cdots}}} \tag{1.2}$$

An equivalent notation, which is more typographically cooperative, is

$$\phi = \frac{1}{1+}\frac{1}{1+}\frac{1}{1+}\frac{1}{1+}\cdots$$

This formula takes the form of an infinite continued fraction. To understand this form, consider a finite version:

$$1+\cfrac{1}{1+\cfrac{1}{1+\cfrac{1}{1}}}=1+\cfrac{1}{1+\cfrac{1}{2}}=1+\frac{2}{3}=\frac{5}{3}$$

Note that in each step of simplification, the most "deeply buried" fraction—for example, 1/1 in the first expression, 1/2 in the second—is the ratio of two consecutive Fibonacci numbers. As the number of "plus ones" and fractions grows, the approximation (1.1) produces equation (1.2).

The second formula that connects ϕ and the number 1 involves nested square roots:

$$\phi=\sqrt{1+\sqrt{1+\sqrt{1+\cdots}}} \tag{1.3}$$

Again, a finite counterpart works out to

$$\begin{aligned}\sqrt{1+\sqrt{1+\sqrt{1+1}}}&=\sqrt{1+\sqrt{1+\sqrt{2}}}\\&\approx\sqrt{1+\sqrt{2.414213562}}\\&\approx\sqrt{2.553773974}\\&\approx 1.598053182\end{aligned}$$

Unlike the situation with continued fractions, the numbers produced by adding 1 and taking square roots do not have a nice structure. To prove Equation 1.3, however, is not difficult. If we let $x=\sqrt{1+\sqrt{1+\sqrt{1+\cdots}}}$, then

$$x^2=1+\sqrt{1+\sqrt{1+\sqrt{1+\cdots}}}=1+x$$

and so $x^2 - x - 1$. Solving this quadratic equation—noting that $x > 0$—forces $x = \phi$.

Representing Numbers Uniquely

How many ways can a number be factored into a product of smaller numbers? Recall that a number is *prime* if it cannot be broken down. The first few primes are 2, 3, 5, 7, 11, and 13. The number 60 can be dissected in 10 different ways (where the factors are listed in nondecreasing order):

$$2 \cdot 30 = 3 \cdot 20 = 4 \cdot 15 = 5 \cdot 12 = 6 \cdot 10 = 2 \cdot 2 \cdot 15 = 2 \cdot 3 \cdot 10$$
$$= 2 \cdot 5 \cdot 6 = 3 \cdot 4 \cdot 5 = 2 \cdot 2 \cdot 3 \cdot 5$$

However, the last product is the sole way which involves only prime numbers. The Fundamental Theorem of Arithmetic claims that each number has a *unique* prime decomposition.

Factoring large numbers is a formidable problem; no efficient procedure for factoring is known. The most challenging numbers to factor are semiprimes, that is, numbers which are the product of two primes. Semiprimes play a huge role in cryptography, the science of making secret codes. This application of huge primes is important enough that the Electronic Frontier Foundation, concerned about Internet security, offers lucrative prize money for producing prime numbers with massive numbers of digits.

The Fundamental Theorem of Arithmetic no longer holds if we replace multiplication with addition. Even a lowly number like 16 cannot be written uniquely as the sum of two primes: $16 = 5 + 11 = 3 + 13$. What if we take only a *subset* of the natural numbers and insist that each number in this set is used at most once? The set of powers of two $\{1, 2, 4, 8, \ldots\}$ does the trick. For example, the number 45 can be written as $45 = 32 + 8 + 4 + 1 = 2^5 + 2^3 + 2^2 + 2^0$. This is equivalent to writing a number in base 2 since 45 in base 2 is simply 101101. Every number has a *unique* binary representation.

Another subset of the natural numbers that gives unique representations involves the Fibonacci numbers. Zeckendorf's Theorem

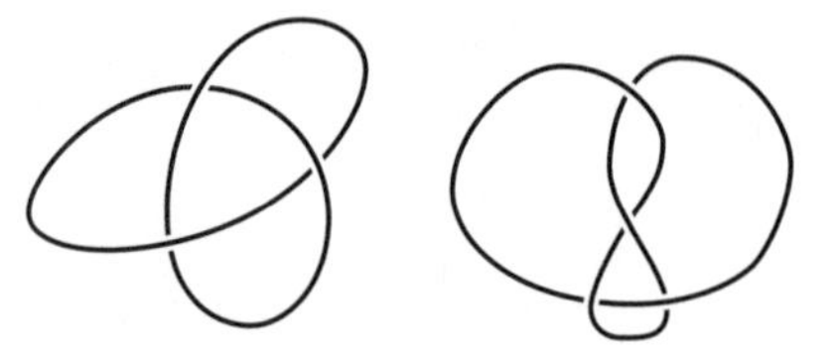

FIGURE 1.2: Trefoil knot (left) and figure-eight knot (right).

states that every positive integer can be represented uniquely as the sum of one or more distinct, nonconsecutive Fibonacci numbers. For example, $45 = 34 + 8 + 3 = F_9 + F_6 + F_4$. Note that we need to insist on having nonconsecutive Fibonacci numbers; otherwise, we could replace F_4 with $F_3 + F_2$ and have another way to represent 45. Although the Fibonacci numbers have been studied for roughly 800 years, Zeckendorf only discovered his result in 1939.

Factoring Knots

In the last section, we saw that every positive integer can be factored uniquely into a product of primes. This idea of decomposing objects into a set of fundamental pieces arises in some surprising contexts.

Knot theory aims to understand the structure of knots—think of lengths of string whose ends are tied together. How does one draw a three-dimensional knot in the plane? Imagine collapsing the knot onto a plane, being mindful of the over- and underpasses. The simplest configuration, a loop of string which makes a circle, isn't really a knot in the conventional sense, and is referred to as the *unknot*. The simplest nontrivial knots are the trefoil (the unique knot with three crossings) and the figure-eight knot (the unique knot with four crossings); see figure 1.2. The figure-eight knot is commonly used in both sailing and rock climbing.

Now comes the idea of decomposition. Imagine taking two nontrivial knots, making a cut in each one, then splicing the two knots together (figure 1.3). We call this a *composite knot*. One could take this process in the opposite direction as well; a knot could be "unspliced" or decomposed into two knots. We are not interested in the case where one (or both) of the new knots is the unknot; this is like saying that

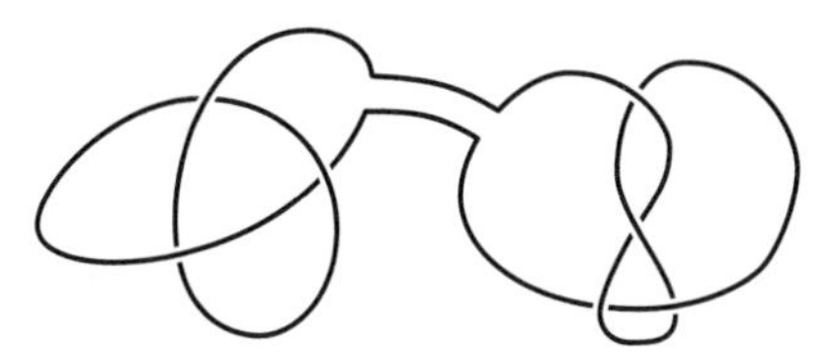

FIGURE 1.3: Adding the trefoil and the figure-eight knots.

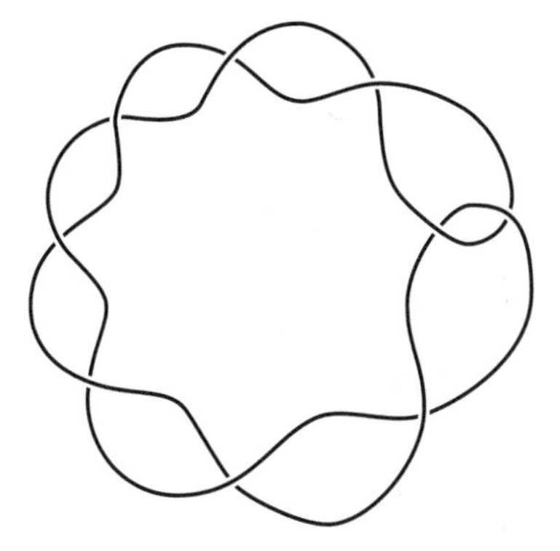

FIGURE 1.4: With one switch of a crossing, can you transform this into the unknot?

a number n can be written as $n \times 1$. If a knot cannot be unspliced, it is called a *prime knot*. The basic question then asks whether every knot can be unspliced into a set of prime knots, that is, does the Fundamental Theorem of Arithmetic extend to the knot setting? Yes! A theorem has shown that every knot has a *unique* prime decomposition. The order of the unsplicing also doesn't matter; regardless of how one unsplices, the end result is always the same set of prime knots.

There are other ways to ascribe complexity to a knot besides counting its crossings. Suppose we cut a knot to switch an overcrossing to an undercrossing (or vice versa). For a given knot, the minimum number of such switches needed to transform it into the unknot is its *unknotting number*. It may seem surprising to learn that there are knots with many crossings which have an unknotting number of 1. Figure 1.4 invites you to switch one crossing to transform this knot with nine crossings into the unknot. Such a situation is typically not recognized with only a casual once-over. Magicians can have fun by taking such knots, making the switch, and watching eyes bulge as a complex mass of crossings melts away to reveal a simple loop of rope. In general, it is relatively difficult to determine the unknotting number of a given knot. However, in 1985 it was shown that if the unknotting number of a knot is 1, then the knot is prime.

Counting and the Stern Sequence

The 19th century mathematician Georg Cantor shocked his contemporaries by developing a hierarchy of different kinds of infinity. Essentially, he came up with a new way to compare the sizes of two sets.

Let's start with a simple problem: How can you show that you have the same number of fingers as toes? Most people would argue, "I have 10 fingers and 10 toes, so they are the same number." This argument is fine, but it brings in an unnecessary concept; the actual *number* of toes and fingers. The question only asked to show that the two sets have the same size, not to actually count them. How else can one answer the question? By making a one-to-one correspondence between the toes and fingers, that is, pair each finger with exactly one toe. The left thumb could be paired with the large toe on the left foot, etc. With these pairings, we claim that the set of fingers has the same size—mathematicians say the same *cardinality*—as the set of toes. This idea of one-to-one correspondence—matching each member of one set to exactly one member of a second set—is what mathematicians formally use to claim that two sets have the same cardinality.

The one-to-one idea goes deeper when one encounters infinite sets. The set of positive integers has the same cardinality as the set of all nonzero integers. How is this possible since the first set fits into the second? Shouldn't the second set be twice as large as the first? Create a correspondence between the two sets:

$$\begin{array}{ccccccc} 1 & 2 & 3 & 4 & 5 & 6 & \cdots \\ \updownarrow & \updownarrow & \updownarrow & \updownarrow & \updownarrow & \updownarrow & \cdots \\ -1 & 1 & -2 & 2 & -3 & 3 & \cdots \end{array}$$

Since every number in the first set matches with a number in the second set, the two sets have the same cardinality. More generally, any infinite

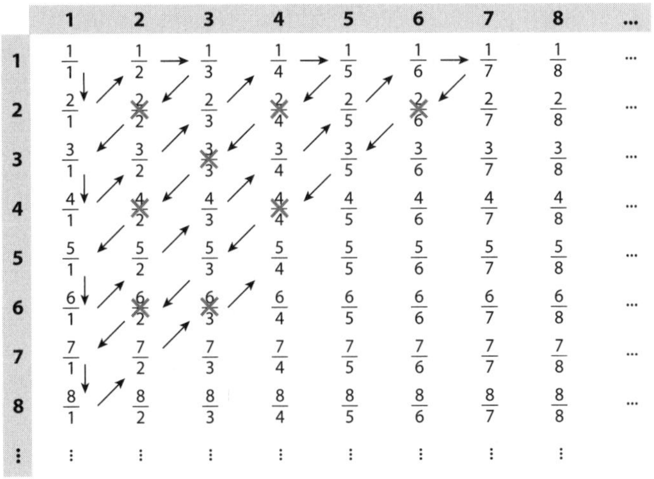

	1	2	3	4	5	6	7	8	...
1	1/1	1/2	1/3	1/4	1/5	1/6	1/7	1/8	...
2	2/1	2/2	2/3	2/4	2/5	2/6	2/7	2/8	...
3	3/1	3/2	3/3	3/4	3/5	3/6	3/7	3/8	...
4	4/1	4/2	4/3	4/4	4/5	4/6	4/7	4/8	...
5	5/1	5/2	5/3	5/4	5/5	5/6	5/7	5/8	...
6	6/1	6/2	6/3	6/4	6/5	6/6	6/7	6/8	...
7	7/1	7/2	7/3	7/4	7/5	7/6	7/7	7/8	...
8	8/1	8/2	8/3	8/4	8/5	8/6	8/7	8/8	...
⋮	⋮	⋮	⋮	⋮	⋮	⋮	⋮	⋮	

FIGURE 1.5: Counting the rational numbers.

set that can be "listed" has the same cardinality as the positive integers. Such sets are called countably infinite, or *countable* for short.

How about comparing the set of positive integers to the set of positive rational numbers? Cantor claimed that these two sets have the same size. Again, how can this be? There are infinitely many rational numbers between any two consecutive integers, making the claim seem ridiculous. The standard approach places the rationals on a grid (figure 1.5) and makes the correspondence by following the diagonal paths. The first few rationals are listed as $1, 2, \frac{1}{2}, \frac{1}{3}, 3, 4, \frac{3}{2}$, and $\frac{2}{3}$. Note from the figure that we have skipped over some numbers. For example, after the number 2/3 is encountered, it reappears as 4/6, 6/9, etc. Taking a first-come, first-served approach, we leapfrog over these latter incarnations of the same number. Each fraction will only be counted when it is in lowest terms.

How can one make the **one-to-one** correspondence *without* doing the skipping? One way involves what is called the *Stern sequence* of integers. This sequence is defined by $f(0) = 0$, $f(1) = 1$, and the two recurrence relations $f(2n) = f(n)$ and $f(2n + 1) = f(n) + f(n + 1)$. The first few terms are $0, 1, 1, 2, 1, 3, 2, 3, 1, 4, 3, 5, 2, 5, 3, 4$. It can be shown that any two neighboring numbers in this sequence are coprime, that is, they share no common factors. This observation eventually

Table 1.1
Stern Sequence.

n	0	1	2	3	4	5	6	7	8	9
$f(n)$	0	1	1	2	1	3	2	3	1	4
$f(n)/f(n+1)$	0	1	1/2	2	1/3	3/2	2/3	3	1/4	4/3

leads to the following remarkable theorem: the sequence of rationals generated by $f(n)/f(n+1)$ generates every positive rational number exactly **once**. We now have the desired correspondence between the positive integers and positive rationals. Table 1.1 shows the first few terms.

While Cantor's ideas about infinity are now considered part of the canon of mathematics, they were a great shock to his contemporaries. Poincaré considered Cantor's work a "grave disease" (Dauben, Georg Cantor, 1979, p. 266), and Kronecker asserted that Cantor was a "corrupter of youth" (Dauben, "Georg Cantor," 1977, p. 89). On the other hand, David Hilbert declared, "No one shall expel us from the Paradise that Cantor has created" (Hilbert, "Über das Unendliche," p. 170).

Fractals

The Ternary Cantor Set is one of the most studied oddball sets in mathematical analysis. To construct it, start by taking the interval $[0, 1]$ and removing its middle third, that is, the interval $[1/3, 2/3]$. Now remove the middle thirds of the remaining two intervals, specifically $[1/9, 2/9]$ and $[7/9, 8/9]$. Keep performing this removal process indefinitely (figure 1.6). It's reasonable to think that nothing will be left at the end of this process. Summing the lengths of the intervals removed, the geometric series helps us see that

$$\frac{1}{3} + 2 \cdot \frac{1}{9} + 4 \cdot \frac{1}{27} + \cdots = 1$$

so the total length removed equals the total length of the interval. However, measuring intervals is not the same as measuring sets of

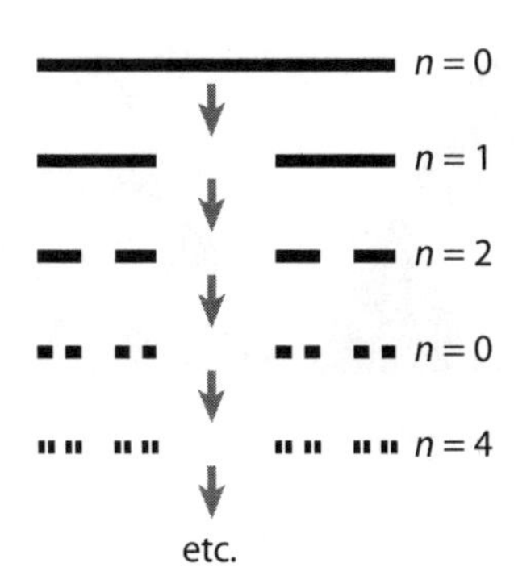

FIGURE 1.6: Approaching the Cantor Set.

points. There are actually infinitely many points left; this is the so-called *Cantor Set*. This set is so fine that it is sometimes called Cantor dust. Without diving into details, this set includes numbers that have an infinite base 3 expansion where none of the digits is a 1. An example is $7/10$, whose base 3 expansion is $0.20022002200\cdots$.

The Cantor Set has another interesting property. Make a copy of the set and scale it by a factor of $1/3$. Make another copy, scale it by a factor of $1/3$, then slide it to the right by $2/3$. The union of these two shrunken sets is exactly the original Cantor Set. When a set can be written as the union of a finite number of shrunken copies of itself, we say it exhibits *self-similarity*. Could we start with another nonempty set, make two copies, scale and shift as before, and get the original set back? No. According to Hutchinson's Theorem, the set of transformations—the rules to shrink and slide copies of a set—*uniquely* determines a set that is self-similar under those transformations.

The process described in generating the Cantor Set can be generalized to build many other strange-looking sets. These sets are more striking when we consider sets of points in the plane. For example, imagine carving an equilateral triangle into four subtriangles and removing the middle one. Now take each of the remaining subtriangles, carve each of them into four pieces, and remove their middle triangles. Continuing this process indefinitely produces the Sierpiński Gasket (figure 1.7).

FIGURE 1.7: Approaching the Sierpiński Gasket.

Self-similarity is also evident with the Sierpiński Gasket: three shrunk copies can be patched together to produce the original figure. While programming the computer to draw a decent approximation of sets like the Sierpiński Gasket is not too taxing, a simpler way is to use the *chaos game*. Consider three possible rules that could be applied to a point in the plane:

1. change (x, y) to $(\frac{x}{2}, \frac{y}{2})$
2. change (x, y) to $(\frac{x}{2} + \frac{1}{2}, \frac{y}{2})$
3. change (x, y) to $(\frac{x}{2} + \frac{1}{4}, \frac{y}{2} + \frac{\sqrt{3}}{4})$

The first rule takes a point and moves it so that it is one half as far from the origin. The second rule does the same but also slides the point to the right by $1/2$. The third rule is also the same as the first rule but also slides the point to the right by $1/4$ and up by $\sqrt{3}/4$.

Each of these rules is inspired by the self-similarity of the set. How does the chaos game work? Pick any point in the plane and randomly apply one of the three rules. Now randomly pick a rule again and apply it to the new point. After repeating this process, say 100 times, start plotting the points (figure 1.8). The Sierpiński Gasket slowly emerges.

In general, suppose one has a finite number of transformations, each of which involves shrinking possibly followed by a shifting and rotation. By starting with any point and applying one of the rules (randomly chosen at each step) many times, Hutchinson's Theorem guarantees that a *unique* set called an *attractor* will be filled up. The structure of the

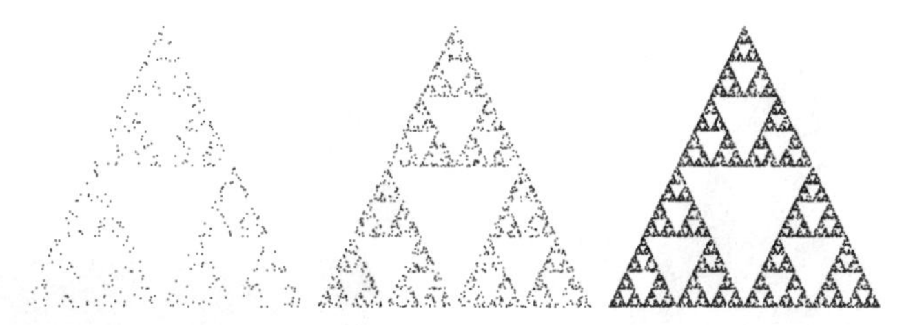

FIGURE 1.8: The chaos game produces the Sierpiński Gasket.

rules guarantees that the attractor will be self-similar. These self-similar sets are called *fractals*. Many intricate sets can be easily generated in this way, including the Barnsley fern and three-dimensional fractals like the Menger sponge (figure 1.9).

Gilbreath's Conjecture

While finding patterns in the primes is something of a Holy Grail, every attempt to find simple order in these numbers has led researchers back to square one—quite literally in the case of Gilbreath's Conjecture. Make a list of the first few primes, then take the absolute value of the differences between successive terms. Then do it again, and again, etc. Table 1.2 contains the first few rows.

Do you notice a pattern? Each row begins with the number 1. And this is not a coincidence because we've only done a few rows. A computer search has verified that the first entry in each row is a 1 for about 3.4×10^{11} rows. Gilbreath's Conjecture states that the first entry will always be a 1. Despite this problem's deceiving simplicity, a single-minded focus is required to crack this conundrum.

Benford's Law

Given a random positive integer, what is the probability that the first digit is a 1? One out of nine, of course. And there's nothing special about the number 1: each of the other digits also has a one out of nine chance of being encountered. What if we change the scenario,

FIGURE 1.9: The Barnsley fern (left) and the Menger sponge (right).

Table 1.2
Gilbreath's Conjecture

2	3	5	7	11	13	17	19	23	29	31
1	2	2	4	2	4	2	4	6	2	6
1	0	2	2	2	2	2	2	4	4	2
1	2	0	0	0	0	0	2	0	2	0
1	2	0	0	0	0	2	2	2	2	0
1	2	0	0	0	2	0	0	0	2	0
1	2	0	0	2	2	0	0	2	2	2
1	2	0	2	0	2	0	2	0	0	0

however, so that we measure population sizes of cities and towns. The likelihood of having a 1 in the first digit is no longer $1/9 \approx 11\%$, but around 30%. And it doesn't apply only to city sizes; income

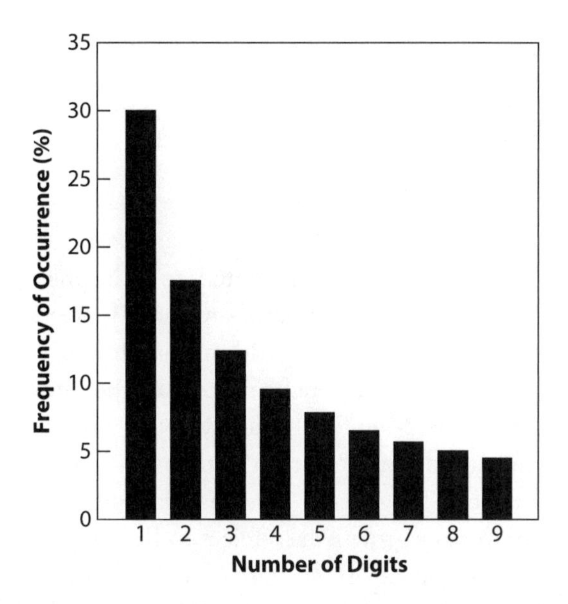

FIGURE 1.10: Distribution of first digits according to Benford's law.

taxes, street addresses, Fibonacci numbers, lengths of rivers, and many other phenomena betray the same bias. To be more specific, Benford's law states that the first digit in these phenomena takes the value n with probability $\log_{10}(1 + 1/n)$. Calculated percentages appear in figure 1.10.

It's easy to see that the sum S of these probabilities equals 1:

$$S = \log_{10}\left(1 + \frac{1}{1}\right) + \log_{10}\left(1 + \frac{1}{2}\right) + \log_{10}\left(1 + \frac{1}{3}\right)$$

$$+ \cdots + \log_{10}\left(1 + \frac{1}{9}\right)$$

$$= \log_{10}(2) + \log_{10}\left(\frac{3}{2}\right) + \log_{10}\left(\frac{4}{3}\right) + \cdots + \log_{10}\left(\frac{10}{9}\right)$$

$$= \log_{10}(2) + \log_{10}(3) - \log_{10}(2) + \log_{10}(4) - \log_{10}(3)$$

$$+ \cdots + \log_{10}(10) - \log_{10}(9)$$

$$= 1$$

The first recorded observation related to this phenomenon was made by the U.S. astronomer Simon Newcomb in 1881. He noticed that the first pages of books of logarithms—such books were needed for calculations—were soiled much more than the later pages. Only in the 1930s did Frank Benford rediscover this observation, and he subsequently measured a large number of data sets, both artificially and naturally constructed. In general, it seems that phenomena that have some kind of power law growth are subject to Benford's law.

This nonintuitive bias of small digits has been used in the service of the law. In attempts to make falsified documents look authentic, tax cheats massage their numbers. In an effort to make their cooked data look real, they manufacture their numbers in a random way. Such violations of Benford's law have triggered audits.

The Brouwer Fixed-Point Theorem

Suppose you have two identical sheets of graph paper. Lay one sheet on a table, but take the other, crumple it up, and lay it on the first sheet so that none of it is hanging over the side. The claim is that there is at least one point on the crumpled paper that is fixed, that is, directly over its "twin" on the flat paper. This is a special instance of the Brouwer Fixed-Point Theorem. Unfortunately, the theorem is not constructive; we have no idea where the fixed point is! Fixed-point theorems like Brouwer's have been used in diverse settings, including mathematical economics. In contrast to the uniqueness theorems witnessed in some of the previous sections, fixed-point theorems are existence results; uniqueness theorems claim that *at most one* of something exists, while existence results assert that *at least one* exists.

A variation of the Brouwer Fixed-Point Theorem is called the Hairy Ball Theorem. Suppose each point on a ball has a short hair pointing outward and the direction of each hair changes in a continuous way. The theorem states that at least one hair must point straight up. You can picture this by trying to comb a coconut. As an aside, there is no such thing as a Hairy Donut Theorem; one could comb all the hairs on a donut flat and in the same direction so that no hair is sticking up.

Inverse Problems

The problem "Given x, find x^2" is a straightforward multiplication problem. Given the number 13, its square is 169. Turning this around, suppose one asks, "Given x, find y such that $x = y^2$." If $x = 169$, then $y = \pm 13$. A host of concerns arise with this second problem. As just witnessed, the answer may not be unique. In fact, a solution may not exist: if x is a negative number, there is no number y whose square equals x. Moreover, even if a solution exists, computing the value (or an approximation) requires much more work; calculating square roots is much more computationally demanding than multiplication. The original squaring problem would be called the *forward problem*, and finding the square root would be the corresponding *inverse problem*.

Inverse problems are usually difficult to solve and analyze. Solutions exist under restricted conditions, and constructing such solutions is usually orders of magnitude more computationally expensive (they require much more computer memory and time) than the corresponding forward problem. Of course, looking for a solution makes no sense if such a solution does not exist, and it may be more difficult to find if the solution is not unique.

Let us consider a more sophisticated inverse problem. Suppose that one could measure the darkness level at each point of a photograph. This information could be used to find the average darkness along any line that intersects the picture. Now "invert" the problem. Suppose the average darkness along any line is known. Can we use this knowledge to find the darkness at each point? The mathematical equivalent of this question was affirmatively answered by Johann Radon in 1917. In fact, the Radon transform is a formula that produces the *unique* darkness levels.

This seemingly theoretical result has enjoyed widespread usage. The first real-world application was seen about 50 years later and concerns medical imaging. Take a cross section of the human body (figuratively, of course). Instead of darkness levels, we are concerned with tissue density at each point. By firing an X-ray beam of known intensity through the body and measuring the reduced intensity when the beam exits, one can calculate the average density of the tissue along that line. Repeat this measurement of average density along all possible lines in the cross section. The Radon transform uses these measurements to reconstruct the density of tissue at each point. By performing this process on stacked cross sections of the body, an image of the body's density at each point is formed. Armed with this information, analysts can detect tumors. Early use of this technique yielded a breakthrough in the diagnosis of neural diseases. In this way, the area of nondestructive medical imaging was born. Alan Cormack and Godfrey Hounsfield won the Nobel Prize in Physiology or Medicine in 1979 for their seminal contributions to this area.

Solving inverse problems has become wildly successful. A key to making the theory work is when the problems have *unique* solutions. Inverse problems also have many applications beyond medicine. Seismic imaging is an inverse problem on a large scale. If a sound wave is blasted into the ground and the variable density of the rock layers is known, one can predict the intensity of the reflected sound waves coming up from the surface. The inverse problem measures the reflected waves and mathematically reconstructs the density of the layers beneath, without digging! This is the modern way of looking for oil. Another inverse problem concerns crack detection. It has become necessary to check engine blocks, cylinder heads, and crankshafts for structural flaws. Similar in spirit to seismic imaging, nondestructive electrostatic measurements reveal the structure of the materials. Essentially, these techniques have proven to be useful in physical situations that value noninvasion. But the story is not over. The theorems behind these applications require *exact measurements*, something impossible to achieve in practice. Limited and noisy data produce images with blurring, streaking, phantoms, and other artifacts.

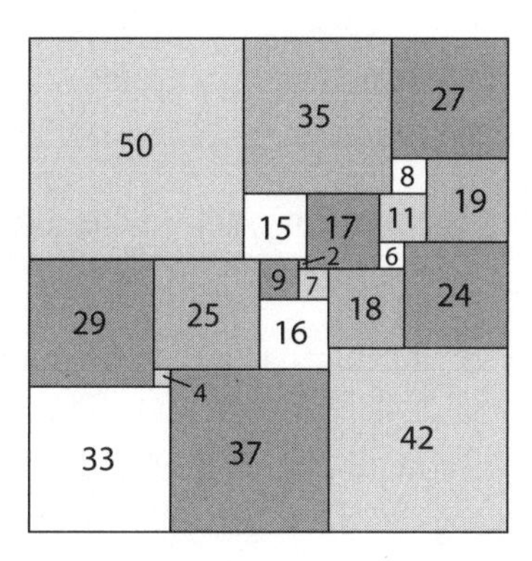

FIGURE 1.11: Duijvestijn's dissection of a square into 21 squares.

More first-class research is needed to develop numerically stable methods to approximate the exact, *unique* solutions.

Perfect Squares

A square that can be dissected into a finite collection of distinct, smaller squares is called a *perfect square*. If no subset of the squares forms a rectangle, then the perfect square is called *simple*.

The Russian mathematician Nikolai Luzin claimed that perfect squares were impossible to construct, but this assertion collapsed when a 55-square perfect square was published by R. Sprague in 1939. In 1978, A.J.W. Duijvestijn delivered a one–two punch by constructing a 21-square simple perfect square (figure 1.11). The number in each square represents the side length of that square. This dissection is *unique* among simple perfect squares of order 21, and there is no simple perfect square of smaller order.

The Bohr–Mollerup Theorem

Students of mathematics usually first encounter the factorial function in conjunction with counting permutations. How many ways can one order the ten letters $\{a, b, c, d, e, f, g, h, i, j\}$? In the first position, there are ten possibilities, leaving nine for the second slot, eight for

Table 1.3
Factorial Function Values

n	0	1	2	3	4	5	6	7	8	9
$n!$	1	1	2	6	24	120	720	5,040	40,320	362880

the third slot, etc. This means that the total number of orderings is $10! = 10 \cdot 9 \cdot 8 \cdot 7 \cdots 3 \cdot 2 \cdot 1 = 3,628,800$. The factorial function is used in many areas of mathematics. Table 1.3 displays how the factorial function grows very quickly. The factorial function can be calculated recursively by using $(n+1)! = (n+1) \times n!$. For reasons connected to combinatorial problems, we define $0! = 1$.

The legendary 18th century Swiss mathematician Leonhard Euler thought about how one could extend the factorial function to the positive real numbers. Using table 1.3, we want to "connect the dots" of the points (0,1), (1,1), (2,2), (3,6), (4,24), etc. Of course, there are infinitely many ways to do this, but we would like a way that forces "nice properties" on the resulting function. Euler defined the gamma function—see figure 1.12—as

$$\Gamma(z) = \int_0^\infty t^{z-1} e^{-t} dt$$

This function enjoys two factorial-like properties: $\Gamma(1) = 1$ and $\Gamma(x+1) = x\Gamma(x)$ for all $x > 0$. These two properties may be combined to show that $\Gamma(n) = (n-1)!$ for all positive integers n:

$$\begin{aligned}
\Gamma(n) &= (n-1)\Gamma(n-1) \\
&= (n-1)(n-2)\Gamma(n-2) \\
&= (n-1)(n-2)(n-3)\cdots 2 \times 1 \cdot \Gamma(1) \\
&= (n-1)!
\end{aligned}$$

Although these two properties restrict the number of possible extensions to the factorial function, there are still many that work. Bohr and Mollerup noted another property satisfied by the gamma function:

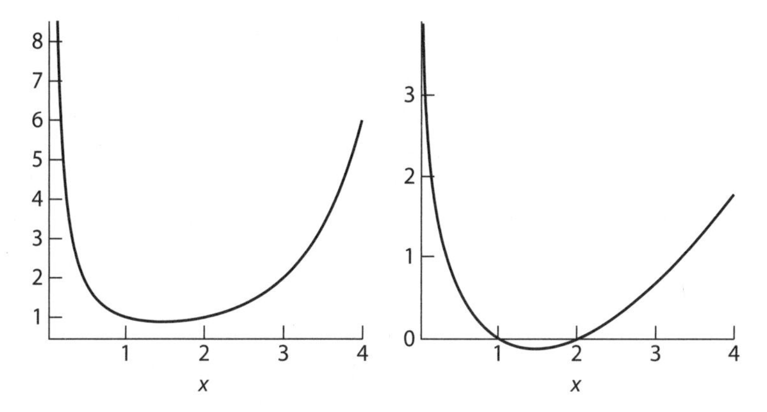

FIGURE 1.12: The functions $\Gamma(x)$ and $\log \Gamma(x)$.

the graph of $\Gamma(x)$ is *log convex*, that is, the function $\log \Gamma(x)$ is convex. To say that a function $f(x)$ is convex means that if $a < b$, then the line segment joining the points $(a, f(a))$ and $(b, f(b))$ lies above the graph of $y = f(x)$. So what makes the gamma function special among all extensions of the factorial function? The Bohr–Mollerup Theorem states that the gamma function is the *unique* function $f(x)$ that is log convex on the positive real axis and that satisfies $f(1) = 1$ and $f(x + 1) = xf(x)$ for all $x > 0$.

The gamma function is widely used in number theory and analysis. Besides the familiar trigonometric functions, for example, sine and cosine, the gamma function is probably the most commonly used special function. Even in statistics, the nonobvious fact that $\Gamma(1/2) = \sqrt{\pi}$ is used in the formula for normal distributions.

The Picard Theorems

When a math student first encounters functions, the concepts of domain (the set of allowable input values) and range (the set of corresponding outputs) are learned. If the domain of a real-valued function is the whole real line, the range can be as large or as small as possible. Simple examples include $f(x) = 5$, whose range is the set $\{5\}$ with just one element, and $f(x) = x$, whose range is the whole real line. What about something in between? The range of $f(x) = 1/(1 + x^2)$ is

$(0, 1]$ (including 1 but excluding 0), while the range of both $\sin(x)$ and $\cos(x)$ equals $[-1, 1]$, and the range of $f(x) = e^x$ is $(0, \infty)$.

Things get more complicated, however, if we extend these functions to having a *complex* variable. Not surprisingly, since the domain has grown, the range generally also grows. Nonconstant polynomials have all complex numbers in their range, a simple consequence of the Fundamental Theorem of Algebra. Both $\sin(z)$ and $\cos(z)$—functions that have a bounded range if the input is real—now also have all complex numbers in their ranges. Some functions that are well-defined on the real line cannot be extended to the whole complex plane, such as $f(z) = 1/(1 + z^2)$, which is not defined at $z = \pm i$.

The Little Picard Theorem makes a broad claim: if a nonconstant function's domain is the whole complex plane and it is differentiable at every point, then the range of the function is the whole complex plane, possibly minus **one** point. For example, the function $f(z) = e^z$ satisfies the conditions of the theorem, and its range contains all complex numbers except for the value 0. So why is this result called the "Little" Picard Theorem? What's so little about it? It's not little, per se; it's simply not as sweeping as a similar result called the Big (or sometimes Great) Picard Theorem. We need more terminology to state this result.

Like functions of a real variable, functions of a complex variable may not be defined at some points. Such points are called *singularities* of a function. Sometimes a singularity is simply a hole in the graph of the function that can be papered over. An example is the function $f(z) = \sin(z)/z$. This function is not defined at $z = 0$, but as z approaches zero, the function's value approaches 1. We call this a *removable singularity*. Another kind of singularity—what mathematicians typically have in mind—is called a *pole*. As an example, $z = 0$ is a pole for the function $1/z$. This is a pole of order 1, while in general a function has a pole of order m at $z = z_0$ if m is the smallest positive integer so that $f(z)(z - z_0)^m$ is either defined or has a removable singularity at $z = z_0$. Finally, we have the scary situation of singularities that are so strong that they are not poles of any order. These are *essential singularities*. An example is $z = 0$ for the function $\exp(1/z)$.

Now what does the Big Picard Theorem claim? Suppose that the function f has an essential singularity at $z = z_0$. Consider a punctured disk centered at z_0, that is, the disk with the center removed. Then the theorem asserts that if we restrict the domain to the punctured disk, the range of the function f is all possible complex values, with *at most one* exception. No matter how small the disk is made, every possible value, with at most one exception, is in the range. The essential singularity $z = 0$ for the function $\exp(1/z)$ misses the value 0 in its range, but it hits every value infinitely many times.

2

The Number Two

Two are better than one.
—Ecclesiastes 4:9

The number 2 is usually associated with pairs or couples. Better than one, we now see how multiple objects relate to one another. This will be seen, for example, with Beatty sequences and the Jordan Curve Theorem. But as we shall see, the number 2 makes beautiful appearances in formulas related to powers and prime numbers. Invest some effort to see the wonderful connections that this number makes. Don't worry; you won't be two-timed.

The Jordan Curve Theorem and Parity Arguments

Mathematicians are assured that once a theorem is proven, it is there to stay. Neither the winds of fashion nor new observations can disprove an established result. That said, what statements can be assumed, and which ones need to be proved? The Jordan Curve Theorem seems so obvious, and yet its proof is quite complicated. The theorem states that any simple, closed curve (or Jordan curve) splits the rest of the plane into two sets, an inside and an outside. Figure 2.1 shows that while some simple closed curves match our intuition of what it means to be simple, others do not. Believe it or not, the Mona Lisa figure is one curve.

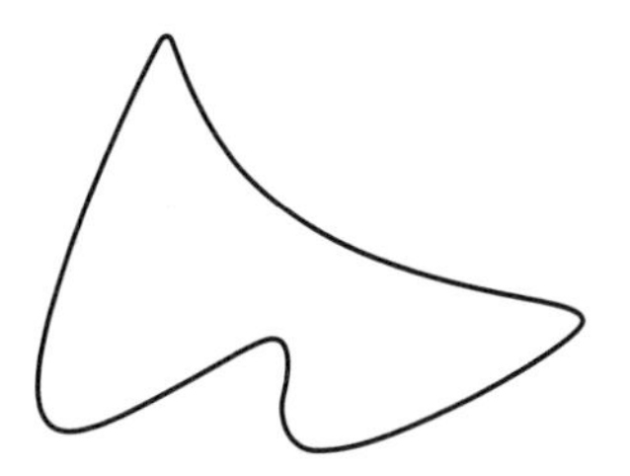

FIGURE 2.1: Jordan curves, simple (above) and not-so-simple (below).

FIGURE 2.2: Is the point inside the curve or outside?

It was constructed by Bob Bosch by cleverly using an algorithm typically used for solving the Traveling Salesman Problem. This example demonstrates that determining whether a point is inside or outside a curve is not always so obvious. Let's find a simple way to do this.

Consider figure 2.2. Starting from the given point, move "outward" until a part of the curve is crossed (we want a transverse crossing, not grazing the curve and staying on the same side).

Crossing the curve takes one from the inside to the outside or vice versa. Continuing in this manner, keep crossing curves until we are clearly outside. By counting the number of crossings, we can determine where we started: an even number of crossings implies we started on the outside, while an odd number of crossings means we started on the inside. This he-loves-me, he-loves-me-not process is an example of a *parity argument*, that is, an explanation where being even or odd figures in.

Another example of a parity argument involves chess. Is it possible to start a knight at one corner of the board and get to the opposite corner by touching every square exactly once in the process? Before you run to get your chess set, consider that each move takes the knight from a white square to a black square or vice versa. Since there are 32 black squares and 32 white squares, a knight's tour must start with one color and end with the opposite color. Since diagonally opposite corners are the same color, the sought-after knight's tour is impossible. More about knight's tours is seen later in the book.

Aspect Ratio

The *aspect ratio* of a rectangle is the ratio of its width to its height. In both videographic and still camera images, a variety of aspect ratios have been utilized. In printed images, a desired property concerns scaling. Suppose we want to bisect a rectangle so that each smaller rectangle has the same aspect ratio as the original. Figure 2.3 illustrates this concept. The practical applications are evident, as one could scale a poster or handbill and make multiple copies on one sheet without distortion. So what is this special value of the aspect ratio?

If the width is x and the height is y, then the halved sheet has the same aspect ratio if $x/y = (y/2)/x$, implying that $x/y = 1/\sqrt{2} \approx 0.707$. Letter size paper commonly used in North America—8.5 in. × 11 in.—has an aspect ratio of 0.773, a bit off. Most other countries use the A-series format. The size A0 is the unique rectangle that has an aspect ratio of $1/\sqrt{2}$ and an area of one square meter. Successive sizes (A1, A2, etc.) are made by halving A0 as described above, thus maintaining the aspect ratio. The size A4 is close to North America's letter size. The advantages of this aspect ratio were already noted in 1786 by the German scientist Georg Christoph Lichtenberg.

How Symmetric Are You?

Symmetry, the geometric notion of pattern and invariance, has long been associated with beauty and form. Its simplest manifestation is *bilateral symmetry*, or mirror-image symmetry. Bilateral symmetry is abundant in the living world. Most animals, including humans, exhibit more or less bilateral symmetry with respect to the sagittal plane, the vertical plane which divides the body into the right and left halves.

There is evidence that some animals prefer more symmetric mates. Facial symmetry in humans has been significantly studied. A popular theory claims that symmetry is viewed as attractive because it is a cue to good health. Indeed, experiments have demonstrated that more symmetrical faces are rated as healthier than less symmetrical faces. Another theory suggests that symmetric faces are perceived as more attractive because symmetric stimuli are more easily processed by the visual system. Yet another theory claims that high facial symmetry in an individual is due to a lack of exposure to stressors during development. In any case, we are biologically wired to value balance when we see it.

Bilateral symmetry is also typically evident in the psychological test known as the Rorschach test (or inkblot test). This test was developed

FIGURE 2.3: Scaled images without distortion.

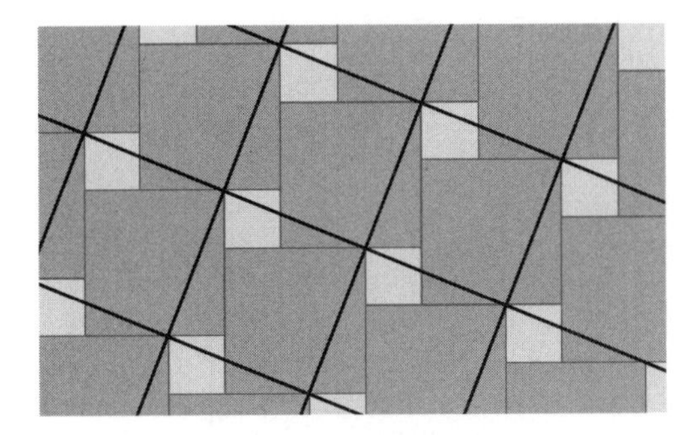

FIGURE 2.4: Pythagorean Theorem proved by tiling the plane with squares of two different sizes.

to understand potential mental disorders in patients who were reluctant or unable to describe their thought processes. Rorschach believed that the symmetry helped to unblock test subjects.

The Pythagorean Theorem

The Pythagorean Theorem is perhaps the most well-known mathematical result in the world. If a and b are the legs of a right triangle and c is its hypotenuse, then $a^2 + b^2 = c^2$. This equation bears the number 2 in its exponents like a beautiful ribbon in a child's hair.

At age 17, budding mathematician Paul Erdős—more on him later—was invited to spend time with the thirteen-year-old son of a prominent businessman in Budapest. The younger boy was showing mathematical promise, so his father wanted to expose him to a rising star. Erdős plied the boy with mathematical questions and facts. Included in their conversation was the following back-and-forth:

"How many proofs of the Pythagorean Theorem do you know?" Erdős asked.

"One," came the answer.

"I know thirty-seven." (Hoffman, *The Man Who Loved*, p. 60)

According to one source, there are 367 proofs of the theorem, including one by Leonardo da Vinci and one by U.S. president James Garfield. One particularly elegant proof that requires no algebra uses a tiling argument. Figure 2.4 shows how the plane can be tiled with squares of two different sizes.

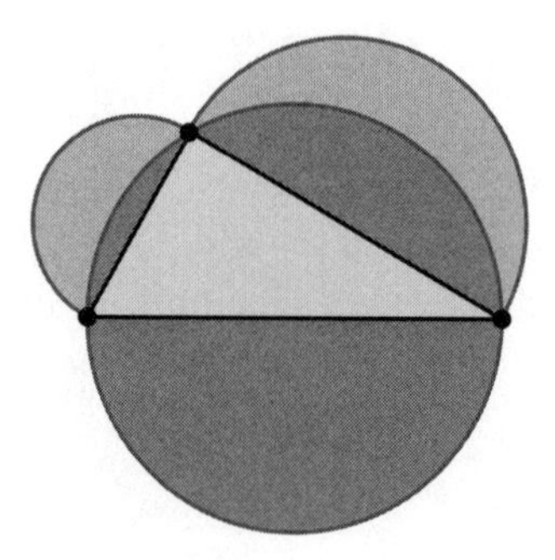

FIGURE 2.5: Does the combined area of the two lunes equal the area of the triangle?

By considering any of the large (tilted) squares in the figure, one finds that it can be decomposed into a medium square and a small square. The sides of the triangles, however, match with the side lengths of the different sized squares, giving us the Pythagorean Theorem. This proof is credited to the ninth century Arabic mathematicians Al-Nayrizi and Thābit ibn Qurra.

Besides its ubiquitous appearances across mathematics, the Pythagorean Theorem is employed by carpenters to ensure that corners are 90 degrees. They refer to this principle as the 3–4–5 rule.

For those who want a challenge, the Pythagorean Theorem can be used to prove a result concerning lunes. A *lune* is shaped like a crescent moon whose boundary consists of two circular arcs. The lunes of Alhazen, dating back roughly 1,000 years and named after a Persian mathematician, is a figure involving two lunes and a triangle; see figure 2.5. Can you prove that the combined area of the two lunes equals the area of the triangle?

To construct right triangles algebraically, the Pythagorean equation is satisfied by

$$(m^2 - n^2)^2 + (2mn)^2 = (m^2 + n^2)^2$$

If m and n are rational numbers, the sides of the right triangle are all rational. The area of such a triangle is $mn(m^2 - n^2)$. A natural question asks what values these areas can take. More specifically, given a positive integer N, is there a rational triangle that has area N? This is called the *Congruent Number Problem*. If such a triangle exists, we say that N is *congruent*.

The right triangle with sides $\{20/3, 3/2, 41/6\}$ has area 5, $\{3, 4, 5\}$ has area 6, and $\{35/12, 24/5, 337/60\}$ has area 7. But it can be shown that rational, right triangles with area $1, 2, 3$, or 4 do not exist. For larger areas, answering the Congruent Number Problem becomes difficult. Because of all the squares involved, one can show that the problem can be simplified to considering only the case when N is a square-free integer, that is, a prime p could divide N but not p^2.

A computationally feasible test known as Tunnell's Theorem produces an answer for a given N. To state this result, we define the following four sets:

$$f(N) = \#\{(x, y, z) \in \mathbb{Z}^3 : x^2 + 2y^2 + 8z^2 = N\},$$

$$g(N) = \#\{(x, y, z) \in \mathbb{Z}^3 : x^2 + 2y^2 + 32z^2 = N\},$$

$$h(N) = \#\{(x, y, z) \in \mathbb{Z}^3 : x^2 + 4y^2 + 8z^2 = N/2\},$$

$$k(N) = \#\{(x, y, z) \in \mathbb{Z}^3 : x^2 + 4y^2 + 32z^2 = N/2\}.$$

This notation means that $f(N)$ is the number of triples (x, y, z), each term being an integer, where $x^2 + 2y^2 + 8z^2 = N$. Tunnell's Theorem applies to square-free N. It asserts that if N is odd, then N is congruent if and only if $f(N) = 2g(N)$, and if N is even, then N is congruent if and only if $h(N) = 2k(N)$. Since each of the four sets is finite for a fixed N, verifying Tunnell's Theorem is a finite calculation that can be checked with a computer. As an example, the number $N = 2$ is not congruent since $h(2) = 2$ and $k(2) = 2$. On the other hand, $N = 5$ is congruent since $f(5) = 0$ and $g(5) = 0$. By 2009, a team of international researchers had checked all values of N under 1 trillion to see which are congruent.

Oh, there's a pesky detail that was omitted in this version of Tunnell's Theorem. An assumption is needed to make this theorem tick: we have to assume that the Birch and Swinnerton-Dyer (BSD) Conjecture is true. Like finding out that the used car you are eyeing does not have a motor, this detail is a deal-breaker. Why? The BSD Conjecture is one of six major unresolved mathematical problems listed as the Millennium Prize Problems by the Clay Mathematics Institute. A solution to each of these problems fetches a prize of \$1 million. While many researchers are energetically pursuing a solution to the

BSD Conjecture, its K2-sized significance suggests that we may be waiting a while.

Beatty Sequences

In chapter 1, we encountered the constant $\phi = (1+\sqrt{5})/2$, the Golden Ratio. Now construct the sequence of integers

$$\lfloor \phi \rfloor, \lfloor 2\phi \rfloor, \lfloor 3\phi \rfloor, \ldots = 1, 3, 4, 6, 8, 9, 11, 12, \ldots$$

The expression $\lfloor x \rfloor$, the *floor* of x, is the greatest integer that does not exceed x. In other words, round down if x is not an integer. The gaps between successive terms are irregular since ϕ is an irrational number. A sequence constructed in this way is called a *Beatty sequence*. If one now stares at the positive integers that are **not** in this sequence, namely

$$2, 5, 7, 10, 13, 15, 18, 20, 23, 26, \ldots,$$

a fascinating observation is made: this new sequence is the Beatty sequence associated with the irrational number $\phi/(\phi - 1)$. In other words, the sequence starting with $\lfloor \phi/(\phi-1) \rfloor = 2$, $\lfloor 2\phi/(\phi-1) \rfloor = 5$, and $\lfloor 3\phi/(\phi-1) \rfloor = 7$ is a Beatty sequence itself. These two sequences are referred to as *Wythoff sequences*. If x is some large integer, roughly $1/\phi$ of the numbers less than x are in the first sequence (we say the first sequence has *density* $1/\phi$) and roughly $1 - 1/\phi$ of the numbers less than x lie in the second sequence. Remarkably, this splitting phenomenon is not unique to the irrational number ϕ. Rayleigh's Theorem (sometimes called Beatty's Theorem) states that for any irrational number $r > 1$, the Beatty sequences formed by r and $r/(r-1)$ generate every positive integer exactly once. That is, every positive irrational number greater than 1 splits the positive integers into two classes, one with density $1/r$ and the other with density $(r-1)/r$.

Results of this type have been taken even farther. It has been shown that Beatty sequences contain infinitely many primes. As an encore to this section, we present the following eye-catcher involving the Fibonacci numbers, the Golden Ratio, an infinite series, and an infinite

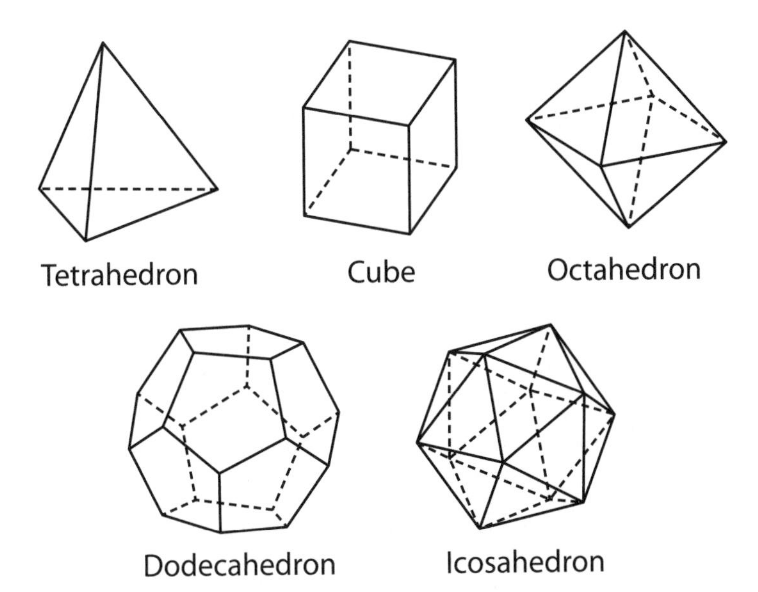

FIGURE 2.6: The Platonic solids.

continued fraction:

$$\frac{1}{2^{\lfloor\phi\rfloor}} + \frac{1}{2^{\lfloor 2\phi\rfloor}} + \frac{1}{2^{\lfloor 3\phi\rfloor}} + \cdots =$$
$$\cfrac{1}{2^0 + \cfrac{1}{2^1 + \cfrac{1}{2^1 + \cfrac{1}{2^2 + \cfrac{1}{2^3 + \cfrac{1}{2^5 + \cfrac{1}{2^8 + \cfrac{1}{2^{13} + \cfrac{1}{2^{21} + \cdots}}}}}}}}}$$

Euler's Formula

In the role-playing game Dungeons and Dragons, players sometimes use five types of dice, each one representing a Platonic solid: the tetrahedron, the cube, the octahedron, the dodecahedron, and the icosahedron. These solids (figure 2.6) have been known since antiquity, dating back to at least 1,000 years before Plato. One may easily count the faces, edges, and vertices of each solid and build table 2.1. Is there a relationship between the number of vertices (V), edges (E), and faces (F) in a Platonic solid? A little exploration with the data in the table reveals that $V - E + F = 2$ in all five cases. Indeed, this relationship holds for *any* convex polyhedron. Discovered around 1750,

Table 2.1

Euler's Formula: $V - E + F = 2$

Name	*Vertices*	*Edges*	*Faces*
Tetrahedron	4	6	4
Cube	8	12	6
Octohedron	6	12	8
Dodecahedron	20	30	12
Icosahedron	12	30	20

this equation is known as *Euler's formula.* It is remarkable that such a simple relationship had been missed for thousands of years. This equation was an inspiration for the development of topology.

Matters of Prime Importance

The number 2 figures prominently in a list of easily stated yet devilishly difficult number theory problems. At the 1912 Fifth Congress of Mathematicians in Cambridge, Edmund Landau mentioned four such problems, which were described as "unattackable" ("Landau's Problems," Wikipedia).

THE GOLDBACH CONJECTURE

Although the primes are the basic multiplicative building blocks of the integers, this didn't stop Goldbach from asking questions about adding primes together. The famous conjecture bearing his name claims that every even integer greater than 2 can be written as the sum of two primes. Posed in 1742, this question remains unsolved, though it has been computationally verified up to 4×10^{18}. A probabilistic argument claims that for large values of n, the expected number of ways to write n as the sum of two primes is approximately $n/(2 \ln^2 n)$. The unboundedness of this function suggests that large values of n can be written as the sum of two primes in many ways. It has been shown that every even number is the sum of at most six primes. In 2013, the related *Odd Goldbach Conjecture* was proved: every odd number above 5 is the sum of three primes. With yet fewer terms, Chen's Theorem asserts that every sufficiently large even number can be written as the sum of a prime and a semiprime.

The Goldbach Conjecture made more than a cameo appearance in some recent fiction. The novel *Uncle Petros and Goldbach's Conjecture* (Doxiadis 2001) concerns a young man, his uncle, and their interaction with some mathematical problems. To gain attention, the publishers offered a $1 million prize to anyone who could prove the conjecture over a two year period. This was a safe bet, and the prize went unclaimed.

One might look for inspiration by considering numbers such as 20 or 38, which cannot be written as a sum of two odd, composite numbers. However, the number 38 is the largest number with this property. You should convince yourself of this claim by examining the following equations:

$$
\begin{aligned}
10k + 0 &= 15 + 5(2k - 3), \\
10k + 2 &= 27 + 5(2k - 5), \\
10k + 4 &= 9 + 5(2k - 1), \\
10k + 6 &= 21 + 5(2k - 3), \text{ and} \\
10k + 8 &= 33 + 5(2k - 5).
\end{aligned}
$$

SQUEEZING PRIMES

A long-established result concerning primes is Bertrand's Postulate: For every $n > 1$, there exists a prime between n and $2n$. This claim was proved by Chebyshev in 1850 by using properties he established concerning the function $\pi(x)$, the number of primes less than or equal to x.

An alternative slick proof could be obtained by using the crown jewel of analytic number theory from the nineteenth century, the Prime Number Theorem. Settled independently in 1896 by Hadamard and de la Vallée-Poussin, this theorem claims that

$$\lim_{x \to \infty} \frac{\pi(x) \ln(x)}{x} = 1,$$

that is, $\pi(x) \approx x/\ln(x)$ for large x. Using this theorem, one has

$$\pi(2n) - \pi(n) \approx \frac{2n}{\ln(2n)} - \frac{n}{\ln n} \approx \frac{n}{\ln n}$$

for large values of n. Since this quantity can be made arbitrarily large, Bertrand's Postulate follows for suffiently large values of n.

Legendre's Conjecture seeks to improve upon this result by finding narrower tracks for primes: for each $n > 1$, there is a prime between n^2 and $(n+1)^2$. Error estimates using the Prime Number Theorem are insufficient to guarantee even one prime in these intervals. This problem is still unresolved.

THE TWIN PRIME CONJECTURE

Euclid proved that there exist infinitely many primes. Dirichlet's sharper result claims that if a and b are coprime, then the set $\{an + b :$ n a positive integer$\}$ contains infinitely many primes. What other sets contain infinitely many primes? Some pairs of primes, such as $(3, 5)$, $(59, 61)$, and $(101, 103)$, differ by two and are referred to as *twin primes*. Like the animals marching into Noah's ark, these twin primes don't seem to stop coming. On Christmas day 2011, a distributed computer project named PrimeGrid announced the largest known twin primes: $3,756,801,695,685 \times 2,666,669 \pm 1$. These numbers have $200,700$ decimal digits.

The Twin Prime Conjecture, still open to this day, asks whether there are infinitely many twin primes. An approach used to tackle this problem concerns infinite series. A classical result shows that the sum $\sum 1/p$ over all primes p diverges. This is a heavy-handed way to show that there are infinitely many primes. Unfortunately, this tactic does not land a one–two blow for twin primes: Brun's Theorem asserts that the infinite series $\sum 1/q$ over all twin primes q converges (whether the Twin Prime Conjecture is true or not). This sum, which is approximately 1.902160583, is called *Brun's constant*.

Curiously, Brun's constant received prime time attention (no pun intended) with its role in shaking up Intel. In 1994, Thomas Nicely, a professor at Lynchburg College in Lynchburg, Virginia, had written some computer code to generate twin primes. When discrepancies appeared between his approximations of Brun's constant on different machines, he eventually was convinced that there was an error in the Pentium CPU he was using. It was soon revealed that rare errors with the floating point division could occur. Almost all the chips produced by that time (more than 1 million) had the same flaw, and though it was relevant in very few applications, the ensuing public relations fiasco pressured Intel to replace all the chips. The company set aside $475 million to cover the expenses.

The Hardy–Littlewood Conjecture, which is stronger than the Twin Prime Conjecture, postulates that the growth of twin primes has a similar form to that of the primes. If $\pi_2(n)$ denotes the number of twin primes less than or equal to x, the conjecture claims that there is a constant C_2 such that

$$\pi_2(x) \approx 2C_2 \frac{x}{(\ln x)^2}$$

for large x. The constant C_2 is defined by the infinite product

$$C_2 = \prod \frac{p(p-2)}{(p-1)^2}$$

where the product is taken over all primes $p > 2$.

A surprising advance related to the Twin Prime Conjecture was made in May 2013. Yitang Zhang (at the University of New Hampshire) showed that there exists some constant $N < 70{,}000{,}000$ such that there are infinitely many prime pairs (p, q) where $p - q = N$. Of course, $N = 2$ is what we would really like, but hey, this is much better than anything else found thus far. Zhang's announcement unleashed a blizzard of activity trying to reduce the value of N. In the summer of 2013, the Polymath8 project announced a new record for N almost every day, sometimes with a minor shaving, sometimes with a scalping. By midsummer, this massive, collaborative effort reduced N to 246.

FIGURE 2.7: How can we cut the shaded area into two equal areas with only one cut?

PRIMES IN A FASTER SEQUENCE

The last conjecture in the cohort concerns how regularly primes are encountered. Dirichlet's Theorem ensures that there are infinitely many primes in any linear sequence of numbers that contain no common factors. What if the sequence grows quadratically? This is the substance of Landau's last question: Are there infinitely many primes of the form $n^2 + 1$? The best result to date shows that there are infinitely many numbers of this form with at most two prime factors.

The Ham Sandwich Theorem

Let's start with a warm-up problem. Grandma baked a 13 in. × 9 in. pan of brownies, which she was planning to divide between her two neighbors. While it was cooling, however, Grandpa took a 3 in. × 3 in. cutting form and cut out a piece for himself; see figure 2.7. Annoyed though she is, grandma still wants to make one vertical cut that slices the remaining brownies into two equal areas. How can this be done? Cut along the line which joins the center of the rectangle with the center of the square. This slice bisects both shapes; hence the area of each piece is the same.

The Ham Sandwich Theorem is a much more general result. Let's double back to where the name comes from. Imagine a sandwich with two irregular top and bottom pieces of bread surrounding a thick slice of ham. Is there one straight cut which halves each slice of bread and the ham? The theorem says that this is always possible. And if you don't like ham, the result is more general. If one has n finite objects in n-dimensional space, there is always an $(n-1)$-dimensional

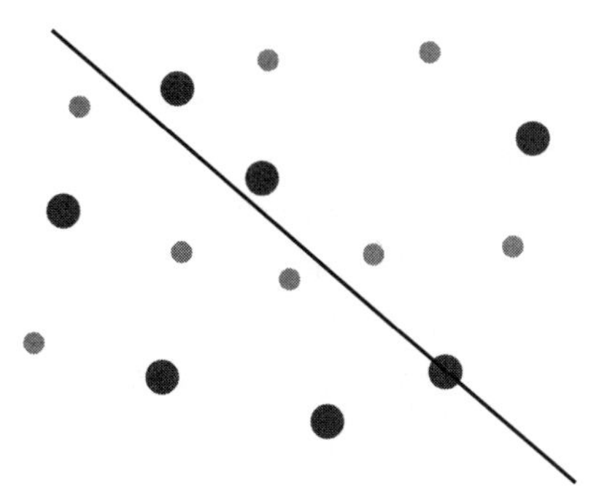

FIGURE 2.8: The line bisects both the set of small and large disks.

hyperplane that simultaneously slices each of the n objects in half. One special note is that each set need not be connected. If one wanted a slice of cheese in the sandwich, think of both pieces of bread as one object and apply the theorem to the ham, cheese, and (combined) bread.

When $n = 2$, an application is sometimes called the *Pancake Theorem*. Two thin pancakes resting on a plate (pretend that they are two-dimensional objects) can be simultaneously cut in half with one straight slice (a line in the plane). Grandma's brownie problem is simply a variation of the Pancake Theorem.

A discrete version of the Pancake Theorem concerns a finite number of points in the plane. Let's represent these as large and small disks (figure 2.8). The theorem claims that there is a cut that leaves half of the large disks on one side of the line and half on the other side. The same cut bisects the set of small disks. If there is an odd number of one type, at least one of the disks is on the line.

Power Sets and Powers of Two

How many subsets can be made from the set $S = \{a, b, c\}$? Since S has only three objects, enumerating the possibilities is not much work:

$$\{\}, \{a\}, \{b\}, \{c\}, \{a, b\}, \{a, c\}, \{b, c\}, \{a, b, c\}$$

So there are eight subsets (note that the first set is the empty set, the set with no elements in it). To answer this question without enumerating the possibilities, one could ask if a given element in S is in the subset under construction. Since there are two possibilities for each element—in or out—and three elements in S, the answer to the original question is $2^3 = 8$. Of course, this can be generalized: if S has n distinct elements, then there are 2^n possible subsets. This collection of subsets of S is denoted by $P(S)$ and is called the *power set* of S.

Powers of two have appeared in many other contexts. The number 2^n equals the number of gaps in the construction of the Cantor Set after n steps. Since 2^n is the number of bit combinations in an n-digit binary number, almost all computer processor registers have sizes that are powers of two (32 or 64 being most common). The classic wheat and chessboard problem depends on 2^n in a dramatic way. There are variations on the background story, but they all culminate with rewarding a wise man with wheat (or in some tellings, rice). One grain is placed on the first square of a chessboard, two on the second, four on the third, etc. Since $1 + 2 + 4 + \cdots + 2^{n-1} = 2^n - 1$ for any n, the total amount paid out to the wise man is $2^{64} - 1 = 18,446,744,073,709,551,615$ grains. This would be a heap of rice larger than Mount Everest and about 1,000 times the global production of rice in 2010. Exponential growth can be dumbfounding.

Powers of two also arise naturally in divide-and-conquer algorithms. These are procedures that break down a problem into multiple pieces (often just two). The procedure is in turn applied recursively to each piece until the problem can be solved directly on small enough pieces. A simple example involves finding a name in a physical phone book (do young people know how to do this?). One first decides whether the name is in the first half or the second half, then one applies the same procedure to the remaining half. This halving process is continued until the set of names is small enough that one can find the name directly. (As an aside, a true story involves a high school math teacher's failure at humor on a calculus test. In some blank space on the test, he pasted in a part of the phone book which included the name "A. Limit," along with the person's phone number. To the teacher's surprise, the students started phoning this person for help in calculus. Mr. Limit was not amused.)

Another example of divide-and-conquer is the Tower of Hanoi problem; see figure 2.9. There are n different-sized disks stacked with decreasing radii on one of three pegs. The goal is to move the whole stack to another peg. There are two important rules in the moving

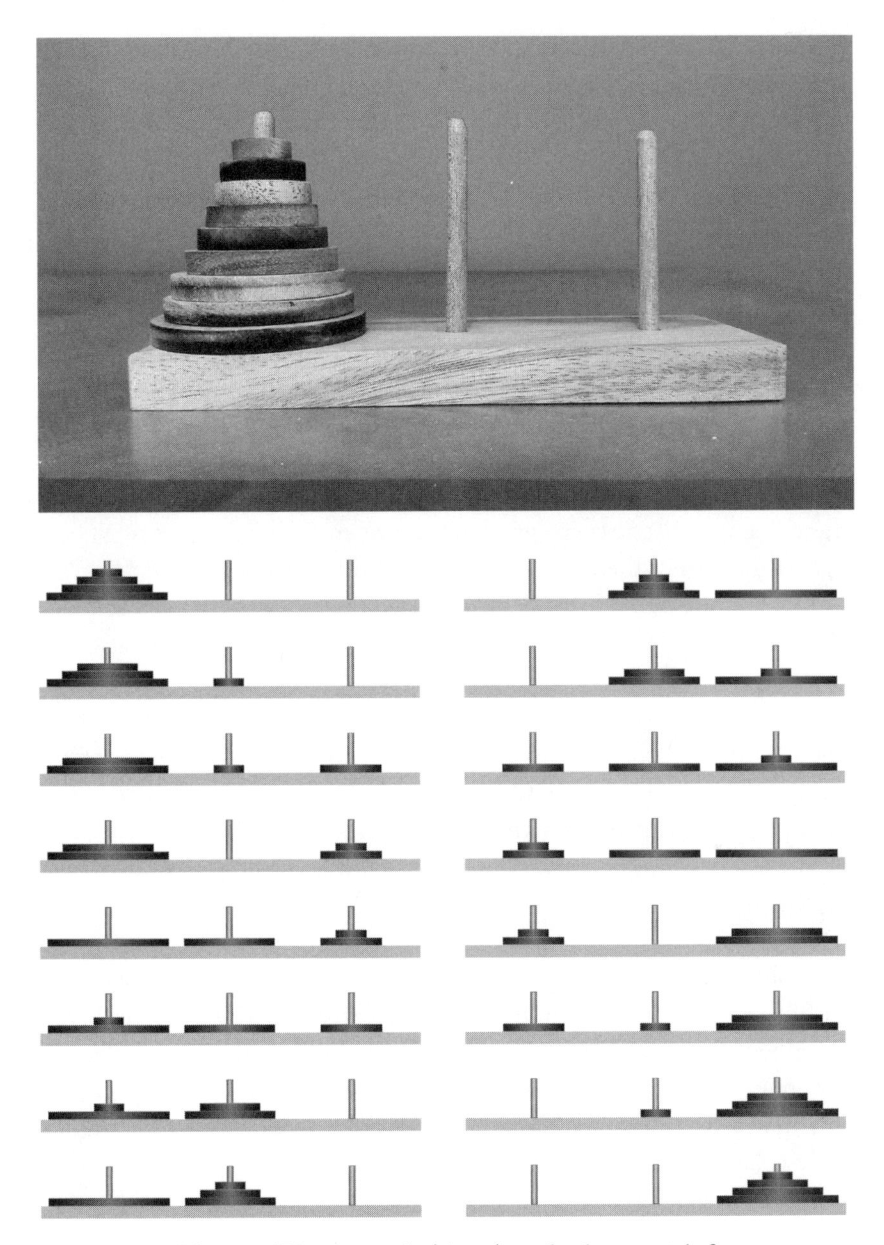

FIGURE 2.9: Tower of Hanoi puzzle (above), and solution with four disks (below).

process: only one disk can be moved at a time, and one cannot place a larger disk on a smaller disk. As the figure reveals, moving the whole stack involves moving smaller stacks. The smallest number of moves needed to solve the Tower of Hanoi with n disks is 2^n-1. Some practical examples of divide-and-conquer include sorting large sets of data (quicksort algorithm) and parsing. Divide-and-conquer algorithms are often the most efficient solving process.

As in the discussion of sets in chapter 1, the question of counting subsets gets more interesting when we move to infinite sets. Of course, if a set is infinite in size, it has an infinite number of subsets. Can one say more? We saw that two infinite sets have the same cardinality (size) if there is a one-to-one matching between their elements. What does it mean to say that the infinite set S_2 is bigger than the infinite set S_1? If there is a one-to-one correspondence between S_1 and a subset of S_2 but not with S_2 itself, then S_2 has a greater cardinality than S_1.

For a finite set S of size n, we saw that $P(S)$ is 2^n. Note that $2^n > n$ for all n. Cantor extended this observation to infinite sets: the cardinality of $P(S)$ is always greater than the cardinality of S. This result immediately implies that the power set of a countable set is uncountable. With more work, one can show that the power set of the natural numbers is one-to-one with the real numbers. Deeper still, Cantor's argument implies that one could keep taking power sets to produce sets of increasingly larger cardinality. Crudely speaking, this means that there are not just two kinds of infinity, but infinitely many kinds of infinity.

Lastly, power sets can be used to show that there are inconsistencies within set theory, at least in the naive sense in which it has been described thus far. Consider making a set S that is the set of all sets. It contains every possible set. Now consider the power set of S. Since this new set has a larger cardinality than that of S, it must be larger, and we have a contradiction. This mind-bending area of mathematical thought attracted a lot of attention in the early twentieth century.

The Sylvester–Gallai Theorem

A set of points in the plane is *collinear* if all the points lie on a common line. If a finite set of points is not collinear, must there be a line which

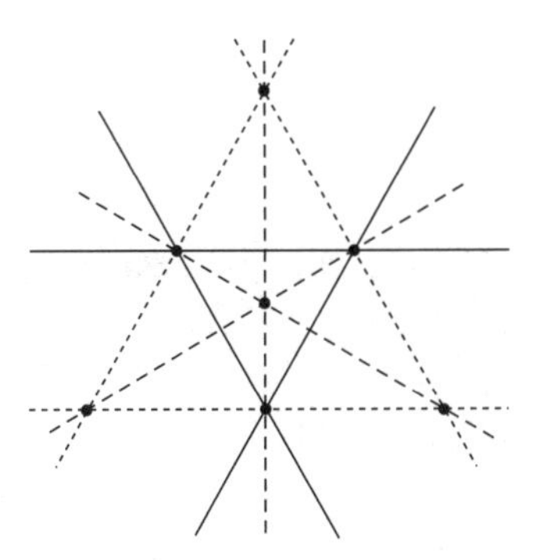

FIGURE 2.10: Exactly three ordinary lines through seven points.

contains *exactly two* of the points (an *ordinary* line)? This question was posed by J. J. Sylvester in 1893. Paul Erdős exhumed the conjecture in 1943, and it was soon solved by his fellow Hungarian Tibor Gallai.

While this problem is easy to state and its claim seems intuitively obvious, formally proving the theorem is a bit slippery. Note however that if an *infinite* number of points is allowed, the theorem no longer holds. Why not? One could place a point at every integer gridpoint in the plane. Every line that goes through two points actually contains infinitely many points.

A more nuanced question asks for the minimum number of ordinary lines $t_2(n)$ that must exist for n noncollinear points. Gabriel Dirac conjectured in 1951 that $t_2(n) \geq \lfloor n/2 \rfloor$. This conjecture is still open, and the best result to date is that $t_2(n) \geq \lceil 6n/13 \rceil$ except for the case when $n = 7$. The example for this case is in figure 2.10.

Formulas for π

The number π is arguably the most recognizable irrational number known to the general public. Memorizing and reciting many digits of π has almost become an Olympic sport. The latest gold medal went to Daniel Tammet, a high-functioning British savant who recited π from memory to 22,514 digits in five hours and nine minutes on Pi Day (March 14), 2004.

There are many formulas used to calculate digits of π, including infinite series, infinite products, and continued fractions. Some of these revolve conspicuously around the number 2. The first is Viète's formula

$$\frac{2}{\pi} = \frac{\sqrt{2}}{2}\frac{\sqrt{2+\sqrt{2}}}{2}\frac{\sqrt{\sqrt{2+\sqrt{2}}}}{2}\cdots$$

Though proven in the sixteenth century, a nice proof follows easily—set $x = \pi/2$—from the infinite product formula found by Euler in the eighteenth century:

$$\frac{\sin(x)}{x} = \cos\left(\frac{x}{2}\right)\cos\left(\frac{x}{4}\right)\cos\left(\frac{x}{8}\right)\cdots$$

A different class of formulas are the so-called Bailey–Borwein–Plouffe, or BBP, series. One such formula, which makes obvious use of powers of two, is

$$\pi = \sum_{n=0}^{\infty}\frac{1}{16^n}\left(\frac{4}{8n+1} - \frac{2}{8n+4} - \frac{1}{8n+5} - \frac{1}{8n+6}\right) \tag{2.1}$$

This formula was only discovered in the 1990s. At face value, it doesn't seem remarkable compared to other formulas for π; some other series representations for π converge much more quickly. The value lies in the disguised observation that this formula can be used to quickly compute digits of π in base 16 without having to use the previous digits. This stands in contrast to most formulas, where each piece needs to be added one at a time. In 2000, 17-year-old Colin Percival used a BBP formula to calculate the quadrillionth binary digit of π in a distributed computer project that required 250 CPU-years and used 1,734 machines in 56 countries.

Multiplication

While some may smirk that multiplication is a topic in this book, this subject is no laughing matter. The number of times (no pun intended) that the product of two numbers is calculated by computers today is truly staggering. Any improvement in the efficiency of the multiplication process can have huge computational ramifications.

Two simple approaches for computing the product of two quantities revolve around the concept of squaring a number. For the first approach, start by computing and storing the values of n^2 for $n = 1, 2, \cdots, 2N$. One can even do this efficiently by using the formula $n^2 = (n-1)^2 + 2n - 1$. Now to multiply any positive integers $x, y \leq N$, use the formula

$$xy = \frac{1}{4}\left((x+y)^2 - (x-y)^2\right)$$

Since addition and subtraction are much less computationally expensive than multiplication, this approach is more efficient than what is taught in school. The division by four may raise some concerns, but keep in mind that if the arithmetic is done in binary, division by four is simply shifting the decimal point by two places. In the nineteenth century, tables of quarter squares had been tabulated up to $200,000$.

The other approach for multiplication that involves squares has a more mechanical feel harkening back to an era of slide rules and other physical devices. By computing squares, one could construct part of the parabola $y = x^2$. Now suppose we want to multiply two positive numbers x_1 and x_2. Mark the points $A = (-x_1, x_1^2)$ and $B = (x_2, x_2^2)$ on the parabola. Simple algebra shows that the line connecting A and B goes through the point $(0, x_1x_2)$. By using string and weights, one can rig up a quick calculator. A model of such a parabolic calculator can be found in the Mathematikum, a German mathematics museum in Giessen; see figure 2.11.

The Thue–Morse Sequence

Playground sports often witness the following team selection process: two captains alternate in picking players for their teams. This system, however, clearly biases the captain who picks first. Is there a fairer way to choose players?

Suppose captains A and B have to choose among eight players. The two fairest ways to order the picking is either ABBABAAB or its opposite BAABABBA. Both of these orderings balance out bias as best as possible. There are similar choices whenever the number of players is a power of 2.

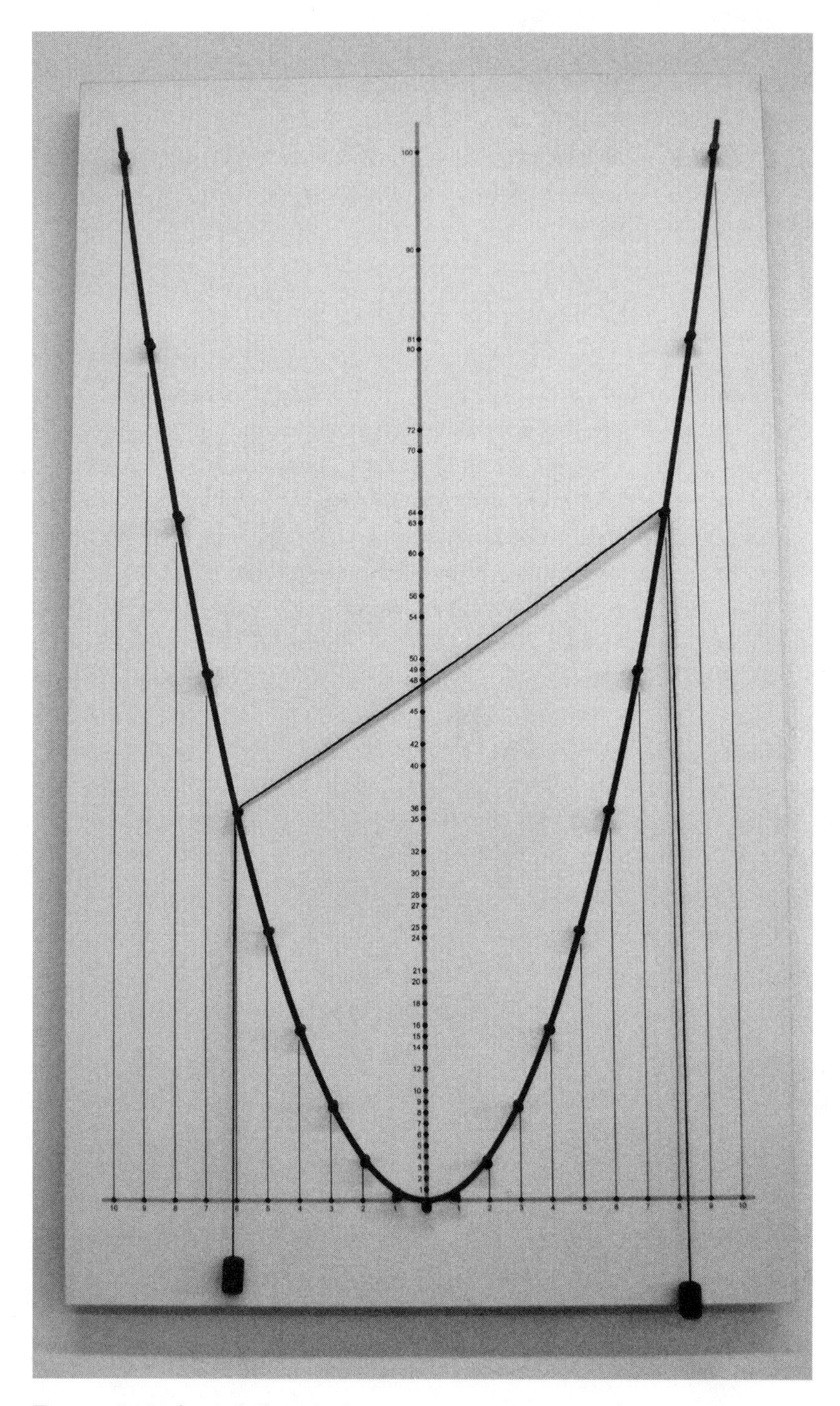

FIGURE 2.11: A parabolic calculator.

These orderings are related to the *Thue–Morse sequence.* This is a sequence of two symbols—we choose zeros and ones—that has various equivalent definitions. The first few terms of the sequence are 01101001100101101001011001101001.

A recursive way of writing t_n, the nth digit of the sequence, is to set $t_0 = 0$, $t_{2n} = t_n$, and $t_{2n+1} = 1 - t_n$ for all n. An explicit form is

$$t_n = \begin{cases} 1, \text{ number of ones in binary expansion of } n \text{ is odd,} \\ 0, \text{ number of ones in binary expansion of } n \text{ is even.} \end{cases}$$

For example, $t_{23} = 0$ since $23 = 10,111_2$. This representation, in terms of odd and even numbers, has inspired the definition of the *odious* numbers as those n for which $t_n = 1$ and the *evil* numbers as those n for which $t_n = 0$:

$$\text{odious}: \ 0, 3, 5, 6, 9, 10, 12, 15, \cdots$$

$$\text{evil}: \ 1, 2, 4, 7, 8, 11, 13, 14, \cdots$$

The Thue–Morse sequence can also be generated by starting with the single digit 0 and at each step simultaneously applying the rules $(0 \to 01)$ and $(1 \to 10)$ to each digit. This means that we replace each 0 with 01 and each 1 with 10. The first few iterations produce

$$0 \to 01 \to 0110 \to 01101001 \to 0110100110010110$$
$$\to 01101001100101101001011001101001 \to \ldots$$

This process of using simple rules to grow an object is sometimes called a Lindenmayer system—L-system for short—and is used in fractal geometry. The Thue–Morse sequence can also be defined as the unique sequence $\{t_n\}$ satisfying the equation

$$\prod_{k=0}^{\infty} \left(1 - x^{2^k}\right) = \sum_{n=0}^{\infty} (-1)^{t_n} x^n$$

The Thue–Morse sequence has other interesting properties. It clearly has a palindromic (mirror-image) structure. Once 2^n terms are formed, make a copy and take its bitwise complement (flip ones and zeros) and append this to get the first 2^{n+1} terms. It is easy to show that the sequence never has three ones or three zeros in a row. This result generalizes in a surprising way. Given any string v of of zeros and ones, the Thue–Morse sequence never has three strings v in a row. In chess, the so-called German rule states that a draw occurs if the same set of moves occurs three times in a row. The chess grandmaster Max Euwe noted that this property of the Thue–Morse sequence allows one to fabricate arbitrarily long chess games.

The Thue–Morse sequence relates to an interesting problem in number theory and ties back to the playground strategy. Using the evil and odious numbers, note that

$$1^0 + 4^0 + 6^0 + 7^0 = 2^0 + 3^0 + 5^0 + 8^0,$$

$$1^1 + 4^1 + 6^1 + 7^1 = 2^1 + 3^1 + 5^1 + 8^1,$$

$$1^2 + 4^2 + 6^2 + 7^2 = 2^2 + 3^2 + 5^2 + 8^2.$$

The "balancing" evidenced here is what implies the fairness in choosing teams. These equations generalize nicely: for any positive integer n, we have

$$\sum_{k=1}^{2^n} (-1)^{t(k)} k^m = 0$$

for $m = 0, 1, 2, \ldots, n-1$. Equations of this type are connected to the so-called Prouhet–Tarry–Escott problem.

Duals

The world of comic hero Superman was jolted with the discovery of a parallel universe—referred to as Bizarro World—where there exists an opposite mirror-image person of each soul on Earth and virtues like beauty are considered vices. As much fun as this was for comic book readers—and this idea was borrowed by the likes of *Star Trek* and

Seinfeld—mathematics has long seen the idea of parallel worlds, though without the emotional baggage and nefarious overtones.

Abstractly, a one-to-one correspondence is made between objects in one space, the *primal* space, to those in another, the *dual* space. The two spaces may look similar or different. The reason for linking these two worlds is that certain properties may seem more apparent in the dual space than in the original primal space.

The concept of pairing quantities or objects from two worlds is more familiar than one may realize. We convert units such as temperature from Fahrenheit to Celsius or distance from kilometers to light years. A more interesting conversion involves the self-proclaimed cyborg Neil Harbisson, who was born with achromatopsia (total color blindness). Since the age of 21, he has worn an electronic eye that converts colors to sound frequencies received by a chip implanted in the back of his head. Harbisson has connected the world of color with the world of sound (and seems to find supermarkets fascinating places to visit).

Examples of duality in mathematics range from simple to fairly sophisticated. The simplest example involves taking positive numbers to negative numbers by multiplying by −1. Note that applying this "flipping" once again brings us back to the original number. When a transformation takes any starting value and returns the same value after two applications of the transformation, we call this an *involution*. Another simple involution is the transformation that takes the reciprocals of nonzero numbers. For example, 5 transforms to 1/5, which transforms back to 5.

A more sophisticated involution is inversion through a circle. Invented by the master geometer Jakob Steiner in 1830, the transformation takes the point (a, b) to $(\frac{a}{\sqrt{a^2+b^2}}, \frac{b}{\sqrt{a^2+b^2}})$. Points inside the unit circle $x^2 + y^2 = 1$ are mapped to the outside and vice versa. But this transformation does a lot more. If we consider how whole sets are changed, one finds that circles which do not intersect the origin transform to circles and that lines transform into circles which intersect the origin. In the 1800s, straight-line guides were needed for precision manufacturing. A mechanical device called the Peaucellier–Lipkin linkage (figure 2.12) was built in 1864 to capture inversion through a circle, thus allowing rotary motion to be converted into straight-line

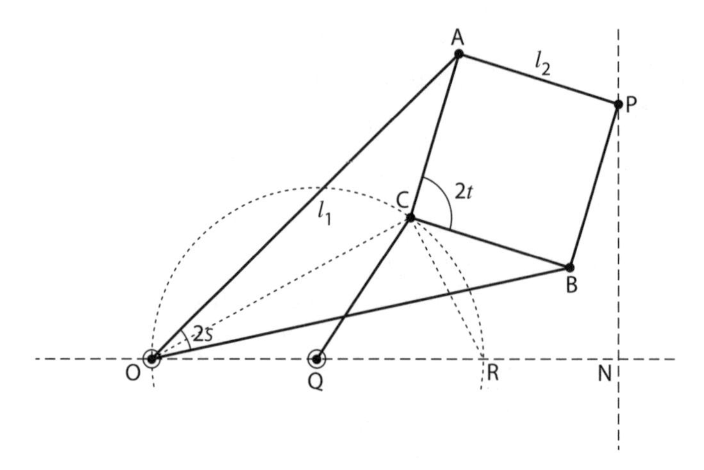

FIGURE 2.12: The Peaucellier–Lipkin linkage.

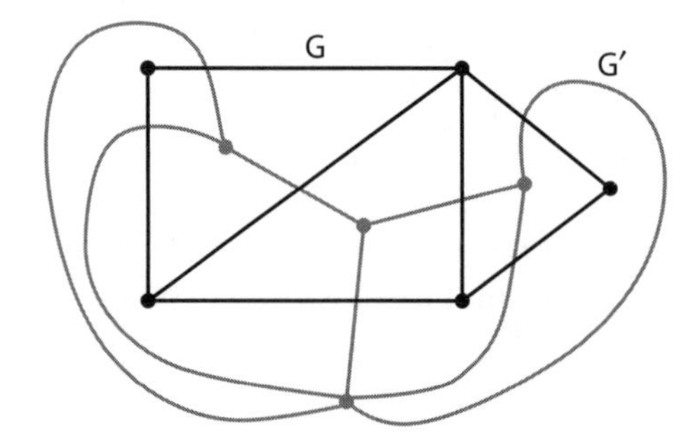

FIGURE 2.13: A graph G and its dual graph G'.

motion. Such tools played an important role in the development of the steam engine.

The last example of involution concerns graphs. Figure 2.13 shows a graph G and its dual graph G'. To construct the dual graph G', place a point in each face of G, including the unbounded region. For each pair of these new points whose faces share an edge of G, connect them with a new edge, which becomes part of G'. The construction of a dual graph is an involution since the dual of G' is simply G. Use figure 2.13 to double-check this yourself.

Let's see how a property in G relates to a property in its dual G'. A graph is *bipartite* if its points can be divided into two sets, say A and B, such that every edge connects a point in A with a point in B.

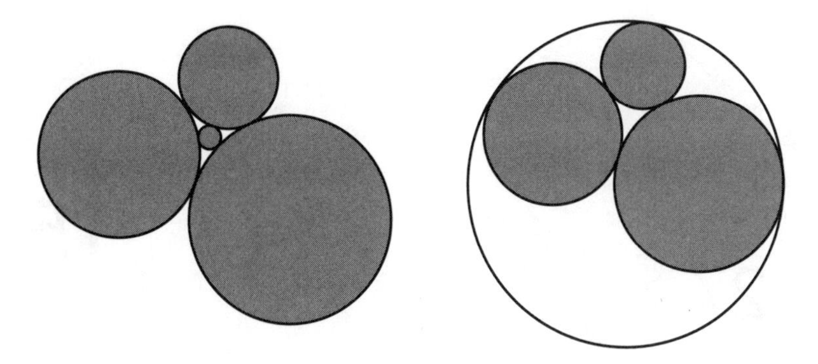

FIGURE 2.14: Two Descartes configurations.

Bipartite graphs arise naturally. An example is the graph whose vertices are organizations and people and the edges connect the organizations to the members they contain (assume that each person is in at least one organization). A graph is said to be *Eulerian*—named after Euler, of course—if it contains a path along the edges that traverses the whole graph. A theorem claims that a graph G is bipartite if and only if the dual graph G' is Eulerian.

Apollonian Circle Packings

We saw in chapter 1 that one can pack a square with different-sized squares. This works with rectangles as well. What about shapes where the sides are curved? This doesn't seem possible, but stretch the idea of packing to allow infinitely many pieces.

Arrange three circles so that they are each touching the other two at exactly one point (some call these “kissing circles”). There are two special circles that can be built in conjunction with this figure: the circle in the trapped space, which is tangent to each circle, and the circumscribed circle, which is also tangent to each of the circles. Either set of four circles is referred to as a *Descartes configuration*; see figure 2.14.

If the radius of a circle is r, the *curvature* of the circle is defined as $c = 1/r$. This makes sense, qualitatively, since for large radii, the circle is large, so the bending is small. Descartes is credited with finding the tantalizing relationship between four circles in a Descartes

FIGURE 2.15: Two Apollonian Circle Packings.

configuration:

$$c_1^2 + c_2^2 + c_3^2 + c_4^2 = \frac{1}{2}(c_1 + c_2 + c_3 + c_4)^2$$

If c_1, c_2, and c_3 are given, Descartes' equation will yield two values for c_4. These values correspond to the radii of the trapped circle and the encompassing circle. The curvature of the outer circle will be negative (this is sometimes called the *oriented curvature*). A little algebra shows that if the curvatures of the initial three circles and the outer circle are all integers, then the curvature of the trapped circle is also an integer. Now the magic really starts. One can continue building trapped circles, and each of them will have an integer curvature. In the limit, infinitely many circles will fill in all the gaps. Filling a circle with infinitely many circles in this way is called an *Apollonian Circle Packing*. Figure 2.15 shows two possibilities. The numbers inside the circles represent the curvatures.

Descartes' equation generalizes to higher dimensions. Instead of using three circles to trap a circle, use four spheres to trap a sphere. In fact, in n-dimensional space, use $n + 1$ hyperspheres to trap another hypersphere. Descartes' equation changes little; the Soddy–Gossett Theorem asserts that $n + 2$ mutually tangent hyperspheres in

n-dimensional space with oriented curvatures $c_j = 1/r_j$ satisfy

$$\sum_{j=1}^{n+2} c_j^2 = \frac{1}{n}\left(\sum_{j=1}^{n+2} c_j\right)^2$$

Perfect Numbers and Mersenne Primes

A number n is *perfect* if the sum of its divisors (including itself) equals $2n$. The first four perfect numbers—6, 28, 496, and 8,128—were known to the Greeks. If $2^p - 1$ is a prime—these are called *Mersenne primes*—then $2^{p-1}(2^p - 1)$ is a perfect number. Euclid conjectured—later proved by Euler—that every even perfect number takes this form. Since it is unknown whether there are infinitely many Mersenne primes, this approach can't be used to show that there are infinitely many perfect numbers.

What exactly is known about Mersenne primes? First note that for $M_p = 2^p - 1$ to be prime, p must also be prime. This holds because if $p = ab$ where $a, b > 1$, then

$$2^p - 1 = 2^{ab} - 1 = (2^a)^b - 1^b$$

The last expression has a factor of $2^a - 1$, so $2^p - 1$ is composite. It's easy to verify that M_p is a prime if $p = 2, 3, 5$, or 7. In case you are wondering, M_{11} is not a Mersenne prime since $M_{11} = 23 \times 89$. M_p is also a prime for $p = 13, 17$, and 19, but it took a long time until Euler found the next Mersenne prime, M_{31}. For higher values of p, the value of M_p grows quickly, so a standard check for primality could take a while. Because M_p has a special form, however, a special test—the Lucas–Lehmer test—was developed to check the primality of Mersenne numbers. With the advent of the computer, much higher values of p could be checked. Computer searches blasted off in the 1990s with the Great Internet Mersenne Prime Search (GIMPS), a distributed computing project where volunteers share their computers to search for Mersenne primes. In 2013, the 48th Mersenne prime was found, $M_{57885161}$, a number with 17,425,169 digits. Because of the efficiency of the Lucas–Lehmer test, the largest known prime at any given time has almost always been a Mersenne prime.

One last fact that connects perfect numbers to the number 2: if n is perfect, then

$$\sum_{d|n} \frac{1}{d} = 2$$

that is, the sum of the reciprocals of the divisors of n equals 2.

Pythagorean Tuning and the Square Root of 2

Pythagoras—besides having his name indelibly stamped on a theorem—is famous for a method of tuning. Unlike his theorem, however, his tuning technique has been replaced and will not be immortalized like his theorem.

If two notes are separated by an octave, the higher note's vibrational frequency is twice that of the lower note. This principle is still used today. The note "halfway" between these is called the *perfect fifth.* Pythagoras decided that the ratio of frequencies between the perfect fifth and the note on which it is built should be 3/2. However, a little mathematical exploration reveals that this ratio creates problems. Let's explore why.

Suppose we wish to tune a piano and middle C is believed to be in pitch. The G above this can be tuned with the 3/2 rule, followed by the D above the G, which has a ratio of $(3/2)^2 = 9/4$ compared to middle C. Jumping down an octave, we therefore have that the D just above middle C has a ratio of 9/8. A few similar steps show that the ratio of E to C is $(9/8)^2$. Continuing this process, we find the ratios for $F\sharp$, $G\sharp$, $A\sharp$, and finally the C above middle C. This produces the absurd claim that $2 = (9/8)^6$. It's not a bad approximation since $(9/8)^6 \approx 2.027$, but one sees that the inaccuracies can compound.

Even without following the arithmetic, it was clear to musicians that Pythagorean tuning was inexact, sometimes peaking eyebrows and offending ears. In the late 1500s, the idea of *equal temperament* was advanced, the proposal that every pair of adjacent notes have the same frequency ratio. With twelve notes in an octave, this common ratio of adjacent notes works out to be $2^{1/12}$. This system is mathematically consistent and has the musical benefit of making each musical piece sound "the same" independent of key. Note that the Pythagorean ratio for a perfect fifth, $3/2 = 1.5$, isn't too far from the equal tempered ratio $2^{7/12} \approx 1.4983$.

It's unclear why Pythagoras did not use this power of two for tuning, but since $2^{7/12}$ is an irrational number, perhaps it is not surprising. Pythagoras and his followers disdained the idea of numbers being irrational. Legend has it that when Hippasus, a Pythagorean philosopher, proved that $\sqrt{2}$ is an irrational number, Pythagoras had him drowned at sea. Terror cautions one not to ponder what Pythagoras would have thought of transcendental numbers!

Among the different ways to show that $\sqrt{2}$ is irrational, the use of a geometric figure is particularly lovely. The proof argues by contradiction. This means that we will assume that $\sqrt{2}$ is a rational number, the opposite of our claim, and eventually reach an absurd statement. This implies that the claim is correct. So suppose that $\sqrt{2} = y/x$ for some positive integers x and y. By the Pythagorean Theorem, we can construct a right triangle whose side lengths are x, x, and y; see figure 2.16. By swinging a circular arc from B to D, we split the hypotenuse into two pieces. The tangent line to the circular arc at D meets the base of the triangle at the point E. The symmetry implies $ED = BE$. We now have a new right triangle CDE, whose side lengths are $y - x$, $y - x$, and $2x - y$. The key is to note that the side lengths of this new triangle must also be positive integers but shorter than the corresponding lengths in triangle ABC. Starting with this smaller triangle, the same process could be repeated again, producing a similar triangle with even smaller integer sides. Obviously this process cannot go on forever, so we have the desired contradiction.

Some mathematicians are rankled by contradiction proofs. For this reason, a constructive proof is sought which shows that there is daylight between $\sqrt{2}$ and any rational number. To start such a proof, suppose that a and b are positive integers. Note that $2b^2$ is divisible by an odd number of twos while an even number of twos divide a^2. This implies that $2b^2$ and a^2 must be different, so $|2b^2 - a^2| \geq 1$. This forces

$$\left|\sqrt{2} - \frac{a}{b}\right| = \frac{|2b^2 - a^2|}{b^2(\sqrt{2} + a/b)} \geq \frac{1}{b^2(\sqrt{2} + a/b)} > 0,$$

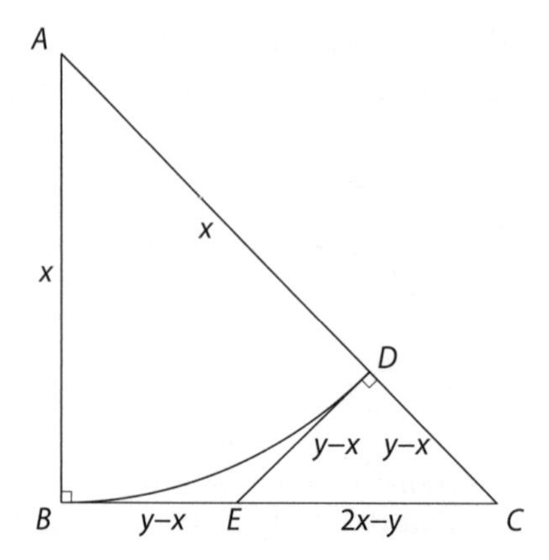

FIGURE 2.16: A geometric proof that $\sqrt{2}$ is irrational.

showing that there is always a gap between $\sqrt{2}$ and any rational number.

Another interesting fact about $\sqrt{2}$: it can be used to show that there exist irrational numbers a and b such that a^b is rational. How? Consider the number $x = \sqrt{2}^{\sqrt{2}}$. If x is a rational number, then we're done. If x is irrational, then

$$x^{\sqrt{2}} = \left(\sqrt{2}^{\sqrt{2}}\right)^{\sqrt{2}} = \sqrt{2}^{(\sqrt{2}\cdot\sqrt{2})} = \sqrt{2}^{2} = 2,$$

so the proof is complete. Lastly, an infinite "tower" formed with $\sqrt{2}$ produces a stunning formula:

$$\sqrt{2}^{\sqrt{2}^{\sqrt{2}^{\cdots}}} = 2$$

Inverse Square Laws

Newton's law of universal gravitation is a succinct but powerful statement:

$$F = G\frac{m_1 m_2}{r^2}$$

This says that the gravitational force attracting two objects whose masses are m_1 and m_2 is proportional to the inverse square of the distance separating the objects. This is not the only physical phenomenon that has an inverse square law. Coulomb's law measures the electrostatic force between two objects:

$$F = k\frac{q_1 q_2}{r^2},$$

where q_1 and q_2 are the charges of the two objects and r is the distance between them. In general, if the intensity of a wave emanating from a point source is uniformly distributed over a spherical wave front, an inverse square law will hold. Why? The surface area of the sphere is proportional to the square of the radius.

The inverse square law for gravitation is not just another pretty face (as far as formulas go). An amazing consequence is Newton's Shell Theorem, which comes in two parts. First, a homogeneous, spherical shell (think of the surface of a hollow ball) gravitationally attracts an external object as if all the mass is concentrated at the center of the shell. A consequence is that a solid sphere whose layers are homogeneous will also attract an external object as if it were a point mass. Planets therefore can be treated as point masses, making calculations much easier. The second part of the Shell Theorem says that an object that is *inside* a homogeneous, spherical shell has no net gravitational effects. It is as if the object is floating in space. This two-pronged Shell Theorem is a potent consequence of Newton's law. In fact, it's so strong that the converse statement holds: if the Shell Theorem is true, then gravity must satisfy an inverse square law.

The Arithmetic-Geometric Mean Inequality

If a movie character needed to scrawl something on a blackboard to demonstrate mathematical prowess, what would he or she write? Most people would imagine an equation like Einstein's iconic formula $E = mc^2$. While fancy equations are considered the markers of genius, often overlooked are inequalities. One of the most used is the *arithmetic-geometric mean* inequality. The number 2 cannot be missed. For any two positive numbers a and b, the arithmetic mean is $(a + b)/2$ and the geometric mean is $\sqrt{ab}$. The inequality claims that the arithmetic mean is at least as large as the geometric mean and that they are equal only when $a = b$. The simplest way to prove this is to observe the equation

$$\frac{a+b}{2} = \sqrt{ab} + \frac{1}{2}\left(\sqrt{a} - \sqrt{b}\right)^2$$

Since the last term is never negative, the inequality follows.

Let's make this inequality work on the double. If $0 < x < y$, let y_1 be the arithmetic mean of x and y and x_1 be the geometric mean. The inequality forces $y > y_1 > x_1 > x > 0$. Now starting with x_1 and y_1, do the same thing. One can continue this process to generate an increasing sequence $\{x_n\}$ and a decreasing sequence $\{y_n\}$. This process can be written succinctly as

$$x_{n+1} = \sqrt{x_n y_n}$$

$$y_{n+1} = \frac{x_n + y_n}{2}$$

One can show that

$$y_{n+1} - x_{n+1} < \frac{1}{2}(y_n - x_n) \tag{2.2}$$

and so the two sequences approach each other quickly. The common limit is called the arithmetic-geometric mean of x and y. Denoting this limit by AGM(x, y), one has the property

$$\mathrm{AGM}(x, y) = \mathrm{AGM}\left(\sqrt{xy}, \frac{x+y}{2}\right)$$

Amazingly, this function relates to the elliptic integral

$$\int_0^{\pi/2} \frac{dt}{\sqrt{x^2\cos^2 t + y^2 \sin^2 t}} = \frac{\pi}{2AGM(x, y)}$$

The AGM function is used to numerically evaluate the integral with a few iterations of x_n and y_n. For example, AGM$(1, \sqrt{2}) \approx 1.198140235$ can seen by calculating a few terms of x_n and y_n; see table 2.2.

Table 2.2
Approaching AGM $(1, \sqrt{2})$

n	x_n	y_n
1	**1.1**892071150027210667174999705604759 1	**1**.2071067811865475244008443621048490 3
2	**1.1981**235214931201226065855718201524 5	**1.1981**569480946342955591721663326624 7
3	**1.198140234**677307205798383788189800 70	**1.1981402347**938772090828788690764074 6
4	**1.19814023473559220743 9**21365592754367	**1.198140234735592207**44063132863310406

The digits in bold represent correct digits. The relation (2.2) can be sharpened by the following equation:

$$y_{n+1} - x_{n+1} = \frac{1}{2}\left(\frac{y_n - x_n}{\sqrt{y_n} + \sqrt{x_n}}\right)^2$$

The square power implies that the number of digits of accuracy roughly doubles with each step, making for lightning fast calculations.

Positive Polynomials

If a calculus student is asked to show that the polynomial $x^4 + 6x^3 + 2x^2 - 34x + 41$ is never negative, he or she is likely to plot the function or use standard techniques about derivatives to find where the function has a minimum. This labor could be circumvented if the student also knew that

$$x^4 + 6x^3 + 2x^2 - 34x + 41 = (x^2 + 3x - 4)^2 + (x - 5)^2$$

Since the polynomial can be written as a sum of squares, it is impossible to make the value negative. But is this technique always possible, that is, can a polynomial $p(x)$ with real coefficients that never takes negative values always be written as a sum of two squares? Yes!

The proof is a nice application of the Fundamental Theorem of Algebra (a polynomial of degree n has exactly n complex roots). Algebraically, we can express a polynomial as a product of linear factors

$$p(x) = C(x - r_1)(x - r_2)\cdots(x - r_n) \qquad (2.3)$$

where $r_1, r_2, \cdots, r_n$ are complex numbers and C is a real number. Since the coefficients in $p(x)$ are real, any complex roots must come in complex conjugate pairs: if $a + ib$ (a, b real) is a root, so is $a - ib$. The product of the two linear factors corresponding to these two roots can be written as

$$(x - a - ib)(x - a + ib) = (x - a)^2 + b^2$$

Grouping all the complex conjugate pairs together allows one to write Equation (2.3) as a product of quadratic terms of the form $(x - a)^2 + b^2$. The Brahmagupta–Fibonacci identity

$$(a^2 + b^2)(c^2 + d^2) = (ac + bd)^2 + (bc - ad)^2$$

can then be used iteratively to combine all these pieces into a sum of two squares. In summary, we have that every nonnegative polynomial with real coefficients can be written as a sum of two squares of polynomials.

This pretty result does not extend to functions with two variables. This fact was known by David Hilbert, but his counterexamples were nonconstructive; he could prove that such functions existed but he did not have a concrete example. Only in 1967 was an example produced by Theodore Motzkin. The function $p(x, y) = 1 - 3x^2y^2 + x^2y^4 + x^4y^2$ (figure 2.17) is nonnegative, but it cannot be written as a sum of any number of squares of polynomials.

Newton's Method for Root Finding

Isaac Newton is famous for his law of universal gravitation, laws of motion, and the development of calculus. Less well known is the method bearing his name for root finding.

The most naive way of finding a root of the function f—a solution to the equation $f(x) = 0$—is the bisection method. This is a divide-and-conquer technique that applies to functions $f(x)$ on intervals $[a, b]$ where f is continuous and the values $f(a)$ and $f(b)$ have opposite signs. The continuity of the function guarantees that the function equals zero somewhere in $[a, b]$. At each step in the process, the interval is bisected, and we know that the function equals zero at some point

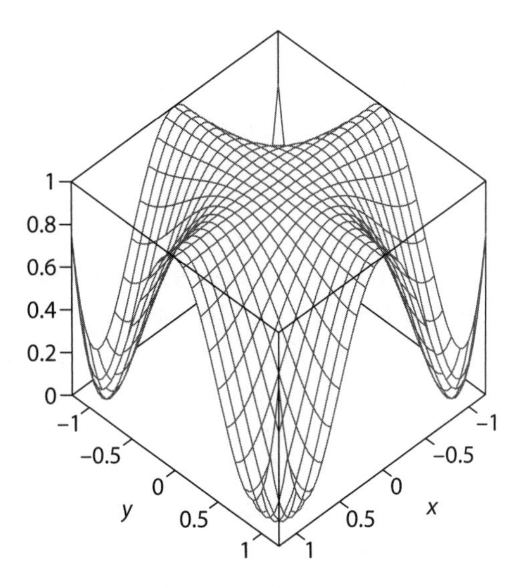

FIGURE 2.17: Graph of the Motzkin polynomial.

in one of the two halved intervals. Iterating this process produces increasingly narrow intervals containing a root of f.

While the bisection method works and is easy to implement on a computer, it is really slow. Each step of the method gains, on average, approximately $\log_{10} 2 \approx 0.3$ decimal digits of accuracy. In contrast, Newton's method is lightning fast. The iterative formula used in Newton's method has a simple, geometrical derivation. Pick a value $x = x_0$ as a starting guess for a root. The tangent line to the curve $y = f(x)$ at the point $(x_0, f(x_0))$ is $y = f(x_0) + f'(x_0)(x - x_0)$. Suppose that this line crosses the x-axis at $x = x_1$. This point becomes our next approximation for the root and satisfies the equation

$$x_1 = x_0 - \frac{f(x_0)}{f'(x_0)}$$

Now starting with x_1, the same process is applied to find x_2, the next guess of the root. Continuing this process, we can write x_n in terms of x_{n-1}:

$$x_n = x_{n-1} - \frac{f(x_{n-1})}{f'(x_{n-1})}$$

So how much quicker is Newton's method than the bisection method? Once we get close to the root, the number of digits of accuracy to the right of the decimal point approximately **doubles** with each step. This makes root finding pinpoint accurate. A caveat is that Newton's method can fail, sometimes spectacularly, if the initial guess is too far away from a root. There are other root-finding methods that yield more digits with each step. Halley's method—named after Edmund Halley of comet fame—triples the number of digits with each iteration. However, the amount of computation needed for each step is no better than two iterations of Newton's method. In many scenarios, Newton's method is the fastest known root-finding technique.

Common examples that use Newton's method involve approximating square roots and higher order roots of polynomials. A more novel application is a fast way to compute the reciprocal of a number. Performing long division is more computationally expensive than multiplication, but Newton's method can be used to approximate the reciprocal of a number by using multiplication. How? To calculate $1/D$, start with the function $f(x) = 1/x - D$. After a little algebra, one sees that the Newton iteration scheme becomes

$$x_n = x_{n-1}(2 - Dx_{n-1})$$

which involves no division!

A natural question related to Newton's method asks about the basin of attraction of each zero. In other words, if $\bar{x}$ is a zero of f, what is the set of initial values x_0 that one can use with Newton's method that will eventually settle down to $\bar{x}$? For example, consider the function $f(x) = x^3 - 1$. There are three roots to this equation: $x = 1$, $x = (-1 + i\sqrt{3})/2$ and $x = (-1 - i\sqrt{3})/2$. Using Newton's method, we find a visually striking image to represent the basins of attraction for each root (figure 2.18). The set of points that approach one of these zeros all have the same shading. The set of boundary points between these different regions is called the *Julia Set*, named after Gaston Julia, a pioneer of complex dynamics. Points on the Julia Set iterate chaotically to other points on the Julia Set.

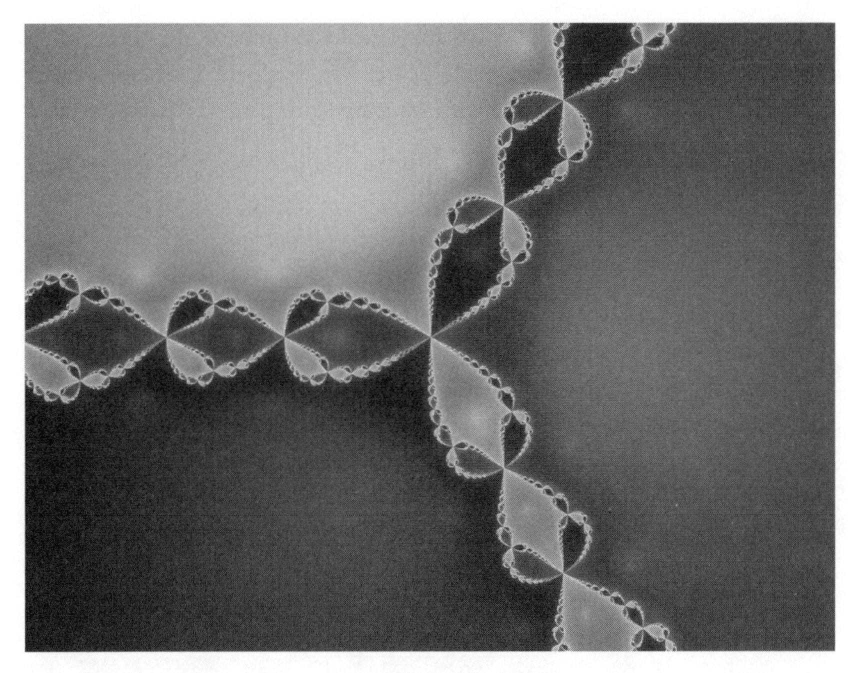

FIGURE 2.18: Basins of attraction for the zeros of $x^3 - 1$.

More Division via Multiplication

We've already encountered the topic of division via Newton's method. That idea showed that division could be replaced with several multiplications. Another technique to accomplish division is similar in spirit:

$$\frac{1}{1-x} = \frac{1+x}{1-x^2} = \frac{(1+x)(1+x^2)}{1-x^4} = \frac{(1+x)(1+x^2)(1+x^4)}{1-x^8} = \cdots$$

If $|x| < 1/2$, then n terms in the numerator approximate the division to a relative error of 2^{-n}. These formulas with powers of 2 can be extended indefinitely to produce the infinite product

$$\prod_{k=0}^{\infty} \left(1 + x^{2^k}\right) = \frac{1}{1-x}$$

for all $|x| < 1$. Another perspective is seen by expanding the infinite product. The coefficient of each power of x equals 1 because every positive integer has a unique, binary representation. This shows that the infinite product equals $1 + x + x^2 + x^3 + \cdots$, the geometric series representation for $1/(1 - x)$.

The Allure of $\pi^2/6$

An unresolved math problem is often named after the person who posed it. Where the Basel Problem—posed as a challenge in 1644—acquires its name is unclear. The Bernoullis, a prominent family of mathematicians from Basel, presumably gave the problem its distinction. But it was another Basel resident, Leonhard Euler, who would solve the problem as a 28-year-old in 1735.

So what's the Basel Problem? It asks for the exact value of the infinite series

$$\frac{1}{1^2} + \frac{1}{2^2} + \frac{1}{3^2} + \cdots \tag{2.4}$$

It's not hard to see that this series must converge by comparing the truncated series to another series:

$$\begin{aligned}
\frac{1}{1^2} + \frac{1}{2^2} + \cdots + \frac{1}{n^2} &< 1 + \frac{1}{1 \cdot 2} + \frac{1}{2 \cdot 3} + \cdots + \frac{1}{(n-1) \cdot n} \\
&= 1 + \left(\frac{1}{1} - \frac{1}{2}\right) + \left(\frac{1}{2} - \frac{1}{3}\right) \\
&\quad + \cdots + \left(\frac{1}{n-1} - \frac{1}{n}\right) \\
&= 2 - \frac{1}{n}
\end{aligned}$$

Note that no matter how large n is, the partial sum is never larger than 2, so the infinite series must converge.

Showing that an infinite series converges is usually not too difficult. Finding the exact value of a converging series is usually a much taller order, if not impossible. There is no obvious reason why the sum (2.4) has a nice closed form. Thus, a double-take is warranted when one encounters Euler's finding:

$$\frac{1}{1^2}+\frac{1}{2^2}+\frac{1}{3^2}+\cdots=\frac{\pi^2}{6} \qquad (2.5)$$

There are now many different proofs of this amazing sum, and they use many different techniques: single integrals, multiple integrals, Fourier series, series representations of trigonometric functions, infinite products, and even number theoretic functions. Euler also observed that his approach could be used for sums where the power 2 is replaced by any even power $2k$:

$$\frac{1}{1^{2k}}+\frac{1}{2^{2k}}+\frac{1}{3^{2k}}+\cdots=R_{2k}\pi^{2k}$$

where R_{2k} is a rational number. And just in case you think that all such problems are solvable, the sum with cubes, namely

$$\frac{1}{1^3}+\frac{1}{2^3}+\frac{1}{3^3}+\cdots$$

does not have a closed form. In fact, only in 1979 was it shown that this number, usually called *Apéry's constant*, is an irrational number.

Equation (2.5) immediately implies that any positive number less than $\pi^2/6$ can be written as an infinite series using only reciprocal squares. This has been extended in a nonobvious way. Suppose that r is a rational number that lies in one of two intervals, $[0, \pi^2/6 - 1]$ or $[1, \pi^2/6]$. Then r can be represented as the **finite** sum of reciprocal squares.

The quantity $\pi^2/6$ arises in another very different situation. If two positive integers are randomly chosen, what is the probability that they are relatively prime? In other words, what is the chance that they share no common factors? To share a common factor implies that a prime factor is shared, so break this down into a simpler problem. What is the probability that two numbers are not both multiples of 2? A random integer has a $1/2$ chance of being even. For both numbers to be even, the probability is $(1/2)^2$, so the odds that the numbers are not both even are $1 - 1/2^2$. Similarly, the probability that two numbers are not both multiples of a prime p is $1 - 1/p^2$. Since this probability is independent between any two primes, we can combine these results: the probability that two random numbers are coprime is given by the infinite product

$$\prod_p \left(1 - \frac{1}{p^2}\right) \tag{2.6}$$

where the product is over all primes p. How are we supposed to evaluate this expression? First, express the terms in expression (2.6) with an infinite series to get

$$\frac{1}{\prod_p \left(1 + \frac{1}{p^2} + \frac{1}{p^4} + \frac{1}{p^6} + \cdots\right)} \tag{2.7}$$

Next, our friend Euler pays us a visit again. Recall that the Fundamental Theorem of Arithmetic says that each positive integer has a unique prime decomposition. Euler realized that expanding the product in expression (2.7) yields the reciprocal squares of all the positive integers:

$$\frac{1}{1 + \frac{1}{2^2} + \frac{1}{3^2} + \frac{1}{4^2} + \cdots}$$

Now we can recycle the solution to the Basel Problem. The probability that two randomly chosen numbers are relatively prime is $6/\pi^2$. Mathematically, Euler plays second fiddle to no one.

Jacobian Conjectures

An old question that has plagued cartographers concerns representing countries accurately on a map. An obvious problem with projecting a spherical design onto a plane is that the wraparound will be violated: France may no longer be next to Spain (or for Risk enthusiasts, Kamchatka is separated from Alaska). While this global concern is avoided if we focus on just a local region, say Central America, we're not out of the woods. All projections introduce some distortion. For example, the popular Mercator projection (figure 2.19) makes

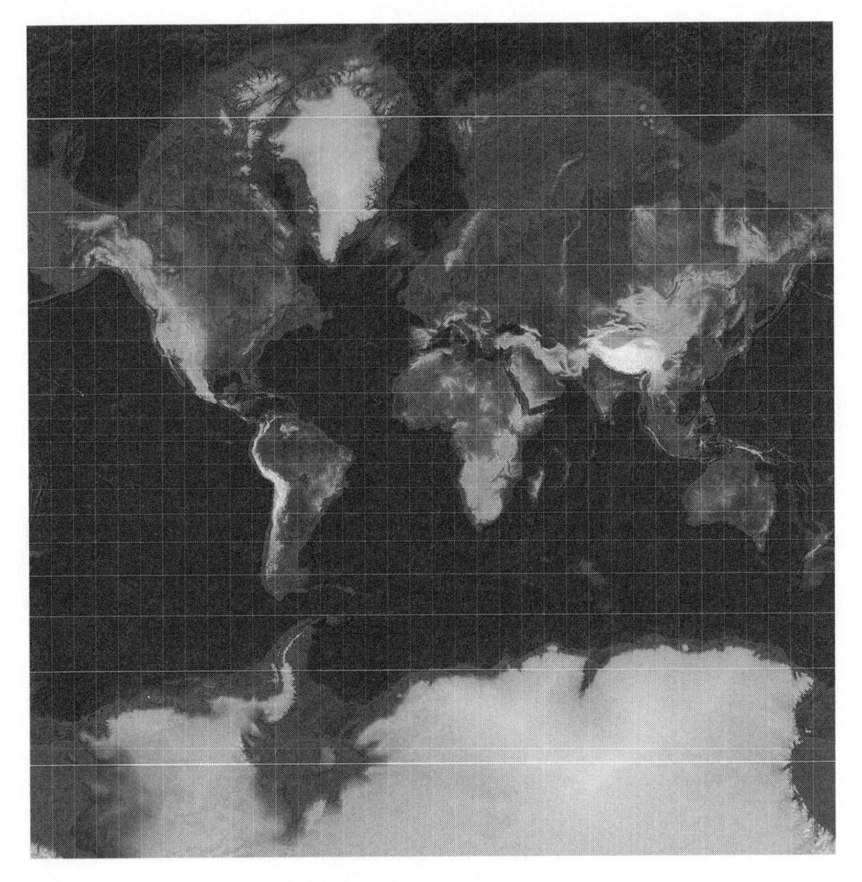

FIGURE 2.19: Mercator projection of the world.

Greenland larger than Africa, whereas in reality Africa's area is 13 times as large. It is reasonable to use a projection that preserves area. Examples include the Mollweide, Sanson–Flamsteed, Hammer, and Eckert IV projections.

What would an equal area projection look like mathematically? Rather than consider the projections of a sphere to a plane, let's consider the simpler phenomena of transformations that transform planar regions into other planar regions. An example is a transformation from the (x, y) plane to the (u, v) plane defined by $u = 2x, v = y/2$. If the (x, y) plane is a rubber sheet, this transformation stretches the sheet horizontally by a factor of 2 but compresses the sheet vertically by the same factor. In this way, the area of any region is left unchanged.

There are many other transformations that preserve area. A more sophisticated example is $u = x + f(y),\ v = y$ where f is any function. Mathematically, the area of any region is unchanged because the determinant of the Jacobian of the transformation equals 1. This condition is equivalent to the equation

$$\left| \frac{\partial u}{\partial x}\frac{\partial v}{\partial y} - \frac{\partial u}{\partial y}\frac{\partial v}{\partial x} \right| = 1 \tag{2.8}$$

A desired property of the transformation is that one could "transform back." Is this always possible? One of the most fundamental theorems in function theory is the Inverse Function Theorem: if Equation (2.8) holds at some point, then the transformation can be inverted near that point. If Equation (2.8) holds on the entire plane, however, it is not clear whether there is a function that inverts the transformation globally. For example, the transformation $u = x + f(y),\ v = y$ can be inverted to give $x = u - f(v),\ y = v$. However, the function

$$u = \sqrt{2}e^{x/2}\cos\left(ye^{-x}\right), \quad v = \sqrt{2}e^{x/2}\sin\left(ye^{-x}\right)$$

is not globally invertible even though the condition in equation (2.8) is satisfied. Both points $(x, y) = (0, 0)$ and $(x, y) = (0, 2\pi)$ map to $(u, v) = (\sqrt{2}, 0)$. Does this mean that Equation (2.8) tells us nothing? Not so fast. Perhaps if we restrict our attention to a certain class of functions, we can obtain the desired global invertibility. In 1939, O. H. Keller proposed that if the transformation has *polynomial* entries, then it is globally invertible and the inverse is also a polynomial. This problem, dubbed the Keller Jacobian Conjecture, is still unresolved. While this conjecture is widely believed to hold in the **two**-dimensional setting discussed, it is suspected to be false in higher dimensions (although a counterexample has yet to be found). The Fields medalist Steven Smale cited the Keller Jacobian Conjecture as one of the most important mathematical problems of the twenty-first century.

3

The Number Three

Three things cannot be long hidden: the sun, the moon, and the truth.
—Buddha

A three-ply cord is not easily severed.
—Ecclesiastes 4:12

The numbers 1 and 2 were tame, ordered, and well-behaved. Clear structure generally held reign. This is not the case with the number 3. With this number, we plunge down the rabbit hole to a world of bouncing numbers, chaotic dynamics, and voting paradoxes. In many scenarios, three represents the number of impossiblity. But rest assured that behind door number 3 you will not find a goat.

The $3x + 1$ Problem

On July 21, 1952, in the southern United Kingdom, Bryan Thwaites had an uphill climb to maintain the attention of a classroom of pupils. After some thought, he came up with a problem that they could work on. Given a small, positive integer, repeatedly apply the following rule:

> If the number is even, divide by two. If it is odd, multiply by three and add one.

Starting with, say, the number 5, the rule produces the sequence $5, 16, 8, 4, 2, 1, 4, 2, 1, 4, 2, 1, \ldots$ The numbers 4, 2, and 1 repeat indefinitely. If we start with 17, this produces the sequence $17, 52, 26, 13, 40, 20, 10, 5, 16, 8, 4, 2, 1, \ldots$ In fact, Thwaites wondered if *any* starting number would eventually lead to the cycle $\{4, 2, 1\}$. It was not obvious. The orbit starting with 27 meanders far and wide; it takes 111 steps to reach 1 (indeed, 96 steps just to get below 27), and the largest number in the orbit is 9,232!

This problem—do all the orbits eventually reach the number 1?—is now usually known as the $3x + 1$ problem. It has also been called the Collatz Conjecture, and to a lesser extent Hasse's algorithm, the Syracuse problem, Kakutani's problem, Ulam's problem, and the hailstone problem, since the problem became popular at certain universities or was popularized by various individuals. In 1960, the Japanese mathematician S. Kakutani said, "For about a month everybody at Yale worked on it, with no result. A similar phenomenon happened when I mentioned it at the University of Chicago. A joke was made that this problem was part of a conspiracy to slow down mathematical research in the U.S." (Lagarias, *The Ultimate Challenge*, p. 32).

Mr. Thwaites notwithstanding, the origin of the problem is generally credited to Lothar Collatz. In 1931, Collatz was considering how complicated number-theoretic functions like the $3x + 1$ rule could produce interesting graphs. Since his initial functions were more complicated, he sought a simpler function that retained the complicated dynamics. The $3x + 1$ problem was born. The Collatz tree—which shows how numbers iterate with this rule—is shown in figure 3.1.

To date, hundreds of research articles and two research monographs related to the $3x + 1$ problem have been published, but it is still unsolved. In 1999, a two-day conference devoted to the problem was held in Germany. Indeed, it can be argued that only minor dents have been made into this problem's impenetrable hull. The problem has been linked to disparate areas of mathematics, including number theory, functional equations, cellular automata, combinatorics, chaos theory, and statistical mechanics, but real progress toward a solution is wanting.

Lest one think that a back-of-the-envelope calculation will reveal a small counterexample to this tenacious problem, think again.

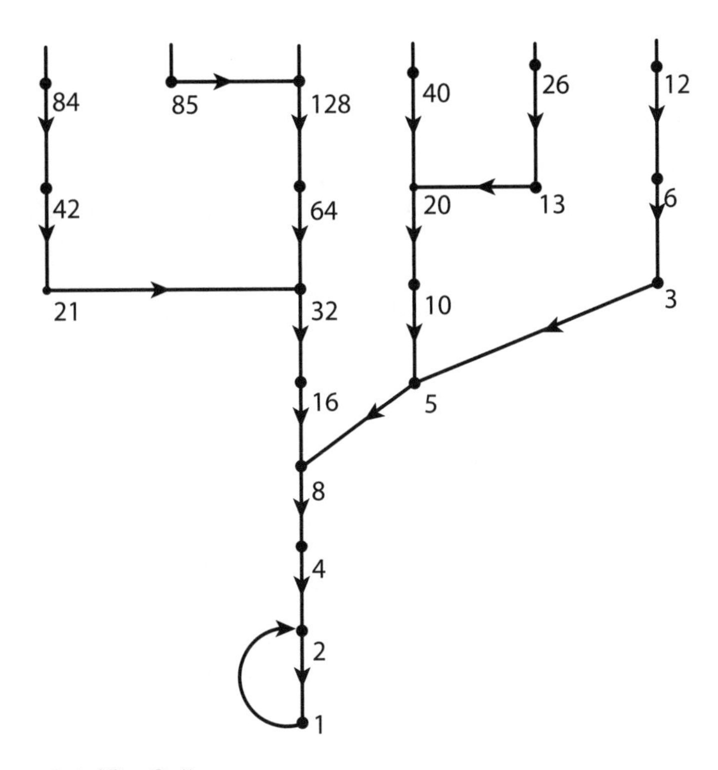

FIGURE 3.1: The Collatz tree.

Computers have verified the conjecture up to 20×2^{58}. If there was another cycle besides $\{4, 2, 1\}$, it has been proved that it must contain at least billions of terms. University of Michigan professor Jeff Lagarias, the undisputed authority on the problem, claims that the problem does not fit into the scope of classical "structural" mathematics. Or as Paul Erdős claimed, "Mathematics is not yet ready for such problems."

Triangular Numbers and Bulgarian Solitaire

How can a number be triangular? Don't get attached to the shapes of digits. A number n is triangular if it takes the form $n = 1 + 2 + 3 + \cdots + k$ for some integer k. The first few such numbers are 1, 3, 6, 10, and 15. The triangle arises if one thinks of stacking rows of coins. Place one coin in the first row and add one more coin for each successive row. The shape of all the objects forms a triangle. A compact form for the kth triangular numbers is $k(k + 1)/2$. A visual proof is in figure 3.2.

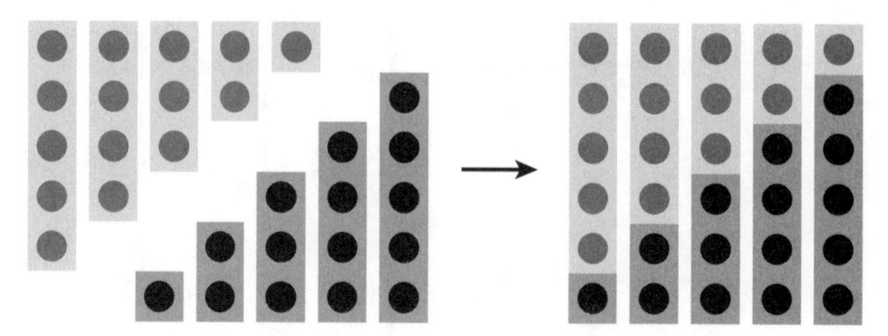

FIGURE 3.2: A proof that $1 + 2 + \cdots + k = k(k + 1)/2$ for $k = 5$.

Triangular numbers appear copiously in combinatorics. Let's see where they fit into a card game. Take a stack of N cards and divide it into several piles, not necessarily the same size. Rearrange the piles so that their sizes are in decreasing order. Now take one card from each pile and collect these into one new pile, again arranging the piles from largest to smallest. Keep repeating this process. This game is called *Bulgarian solitaire*. As an example, suppose that $N = 10$ and the initial piles have size $(8, 1, 1)$. An iteration leads to the piles $(7, 3)$, followed by $(6, 2, 2)$, $(5, 3, 1, 1)$, $(4, 4, 2)$, $(3, 3, 3, 1)$, $(4, 2, 2, 2)$, $(4, 3, 1, 1, 1)$, $(5, 3, 2)$, $(4, 3, 2, 1)$, and then the pattern remains the same.

Since the total number of cards in any game stays fixed, the pattern needs to repeat at some stage. The example above is a special case of a theorem that says that if the number of cards is a triangular number $N = 1 + 2 + \cdots + n$, then any initial configuration will eventually settle into the pattern $(n, n-1, \ldots, 3, 2, 1)$. If N is not a triangular number, then the patterns never reach a fixed state but cycle. For example, starting with $N = 9$, the initial configuration $(5, 3, 1)$ iterates to $(4, 3, 2)$, $(3, 3, 2, 1)$, $(4, 2, 2, 1)$, $(4, 3, 1, 1)$, and back to $(4, 3, 2)$. Someone posed the conjecture that for a fixed N, any initial configuration eventually leads to a *unique* cycle. However, the conjecture is false. In the $N = 8$ case, there are two different cycles:

$$(3, 3, 1, 1), (4, 2, 2)$$

and

$$(3, 2, 2, 1), (4, 2, 1, 1), (4, 3, 1), (3, 3, 2)$$

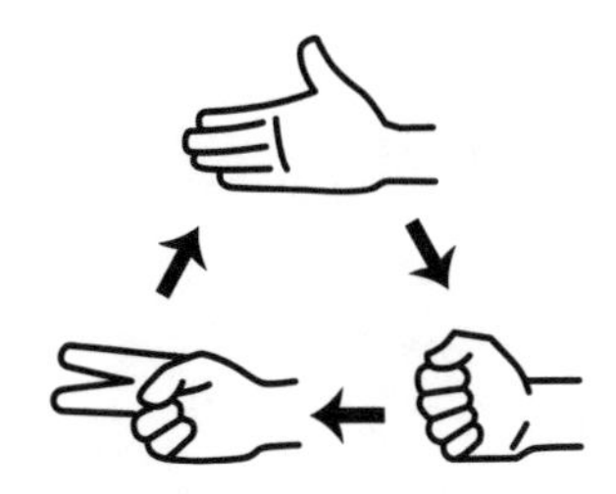

FIGURE 3.3: Rock-paper-scissors dynamics.

Rock-Paper-Scissors and Borromean Rings

While coin tosses, drawing straws, and rolling dice are all well-worn methods for generating random numbers, the game rock-paper-scissors is more recent, and with some skill, not truly random. If you are not familiar with this two-player game, it's quick to grasp. At the same instant, each player puts his or her hand in one of three positions: rock (fist), paper (flat palm), or scissors (fist with the index and middle fingers sticking out). The winner of the game is decided as follows: paper covers rock, rock smashes scissors, and scissors cut paper. This is illustrated in figure 3.3.

While the rules suggest that the game is random, an experienced player can usually win. Indeed, computer programming contests are held where participants use the case histories of their opponents to develop better strategies. Even without computers, tricks can improve one's play. For example, it has been observed that inexperienced men typically start with rock and inexperienced women usually start with paper. The psychology can be intriguing.

Mathematically, this game is of interest because it contains *intransitivity*. Each position beats another position and loses to another position. This aspect connects rock-paper-scissors with a seemingly disparate object: Borromean rings. As we saw in chapter 1, knots can get, well, rather knotty. A set of distinct knots that are connected is called a *link*. Even more confusing, a chain is a simple example of a link; it is composed of several unknots that are linked together. Of course, a length of chain has the property that if an interior link is removed, the chain is decomposed into two parts. A basic question asks whether it is possible to have a link where *all* the pieces split apart if any one knot is removed. The answer is affirmative; these are called *Brunnian links*.

FIGURE 3.4: Borromean rings.

The simplest example of a Brunnian link is the Borromean rings (figure 3.4).

Although the Borromean rings consist of three unknots, these rings are not thickened circular unknots lying in a plane. Why not? Note that ring 1 lies on top of ring 2, ring 2 on top of ring 3, and ring 3 on top of ring 1. Since each ring is above another ring but below the remaining ring, the same intransitivity is evident here as was seen in rock-paper-scissors.

Versions of the Borromean rings have appeared in diverse scenarios, from a symbol used in religion (Buddhist and Hindu temples and the Christian trinity) to company logos (figure 3.5). The name comes from their use in the coat of arms of the aristocratic Borromeo family of northern Italy. The Borromean rings are literally tied to one ancient pattern: braids. Joining the ends of one iteration of the standard "pigtail" braid produces the Borromean rings (figure 3.6).

Random Walks

Five-year-old Monica lives on a street with a long sidewalk. One day, standing in front of her house, she decides to try an experiment.

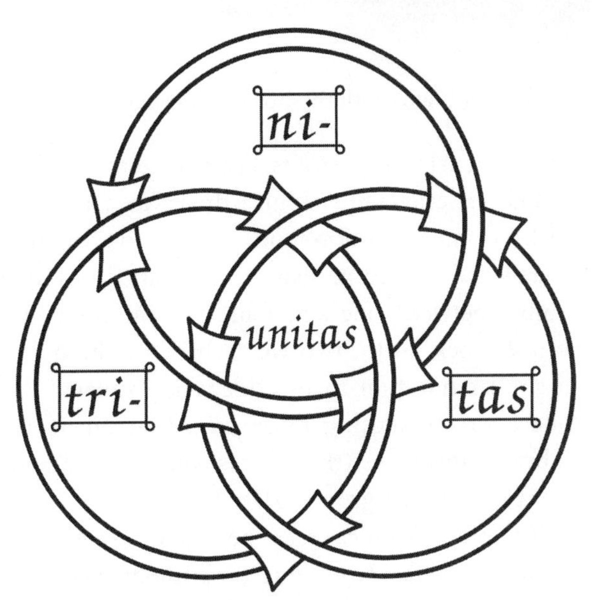

FIGURE 3.5: Borromean rings in a coat of arms (upper left), a beer logo (upper right), and a symbol for the Christian trinity (lower).

She flips a coin and decides that if it lands on heads, she will walk one step north, while if it lands on tails, she will walk one step south. Monica wonders whether she will always eventually return back to the same spot in front of her house. After investigating this many times, it seems that she always returns, although sometimes it takes many steps.

FIGURE 3.6: Connecting braids with the Borromean rings.

Fifteen years later, Monica recruits her younger sister Adèle in a more sophisticated experiement. Monica drives to a part of the city where the blocks form a uniform grid pattern. At an intersection, Adèle rolls a four-sided die (in the shape of a tetrahedron) to randomly generate a number between one and four. She interprets a roll of 1 to mean drive one block north, two west, three south, and four east. Monica has a similar question to the sidewalk problem: Would she always arrive back to the starting intersection? Again, it seems that the answer is always affirmative (though gas consumption and a minor fender bender left their father growling).

Many decades later, Monica tells her great-grandsons, JJ and Quincy, about the sidewalk and driving experiments. With a wink in their eyes, JJ and Quincy knew that they had to take their spacecraft and try something similar. JJ pilots the ship to a fixed point in the ionosphere. Quincy rolls a six-sided die to determine whether the ship should travel 1 kilometer away from the planet, toward the planet, or in one of the four lateral directions. Will the ship always return to the starting position? Repeated experimentation yields a surprise; they seem to return to the starting point only about 1/3 of the time. JJ and Quincy fly home, wondering whether their great-grandma was spinning another tall tale.

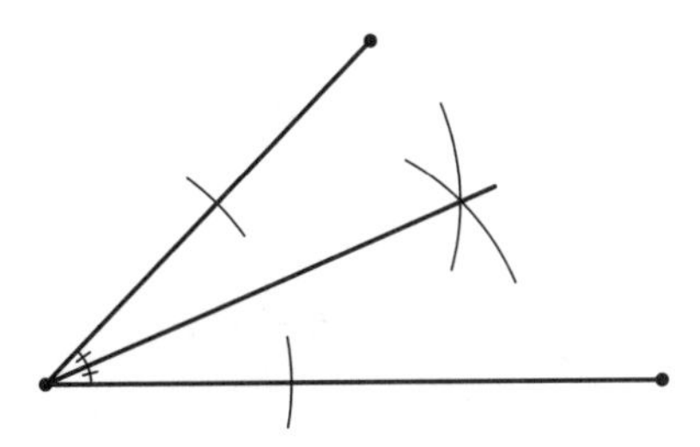

FIGURE 3.7: Can you bisect an angle with a straight edge and a compass?

These experiments are formally known as *random walks*. This type of movement with random steps has been used in many areas of research, such as the movement of molecules in liquids and gases, the phenomenon of genetic drift in population genetics, and in computer science to estimate the size of the Web. The experiments of Monica and her family demonstrate that there are different outcomes for different-sized spaces. Random walks on infinite lattices in one and two dimensions have the surprising property that starting at one point, one will almost surely return. The three-dimensional lattice offers a surprise; the probability of return is approximately 34%.

Trisecting an Angle

Geometry is an ancient area of study that was embraced by many ancient cultures—Babylonian, Egyptian, Indian—but perhaps its greatest legacy comes from Greece. Euclid, Pythagoras, Thales, and Ptolemy (to name a few) left theorems and problems that generate research activity to this day.

One of the problems posed by the Greeks, and for which they obtained no resolution, is the angle trisection problem. How can one trisect a given general angle with only an unmarked straight edge and a compass? Note that angle *bisection* is easy; see figure 3.7.

Trisecting an angle is not as obvious. Special angles—right angles, for example—can be trisected, but the Greeks could not trisect a general angle. It was not until 1837 when Pierre Wantzel used abstract algebra—specifically Galois theory—to show that trisecting a general angle, using only the aforementioned tools, is impossible.

The trisection problem *can* be solved if other tools are allowed. Techniques have been found by using paper (origami), a marked ruler,

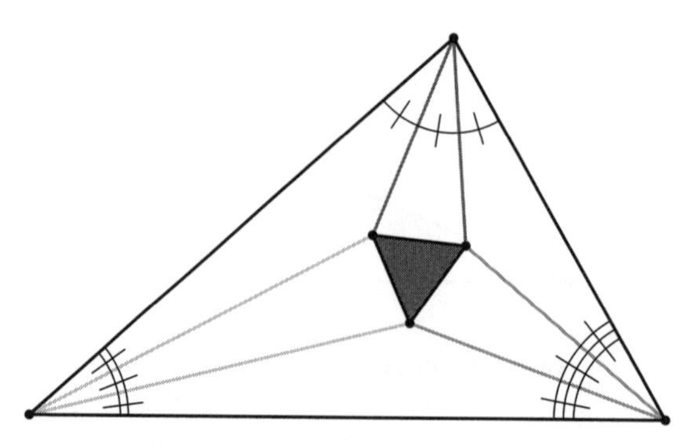

FIGURE 3.8: Morley's Miracle.

or a string wrapped around a cylinder. The simplicity of the original problem and its resistance inspired many amateurs over the years to take their turn at finding an elementary solution. Some also did not accept the nineteenth century result and produced "proofs" that a general angle *can* be trisected. Details are captured in the entertaining book by mathematician Underwood Dudley, *The Trisectors* (1996).

Though we can't trisect an angle, this has not impeded the discovery of theorems where trisections are involved. A gorgeous result known as Morley's Trisector Theorem claims that for any triangle, the points of intersection made by the lines that trisect an angle form an equilateral triangle (figure 3.8).

This result, found in 1899, inspired such awe that it has also been called Morley's Miracle.

The Three-Body Problem

Newton's law of universal gravitation has many different applications. While it has seen spectacular success in ballistics, we want to think on a larger scale where large bodies such as planets and stars come into play.

Some situations with many bodies can be approximated by combining several two-body problems together. When one body is much more massive than the others—as in the case with the sun and the other planets—the motion of each small body is dictated principally by the massive body. From a gravitational perspective, only the sun exists as far as one of its orbiting planets is concerned. A beautiful consequence of Newton's law is that it mathematically confirms the observational results known as Kepler's three laws of planetary motion. Another common two-body problem involves binary stars, that is, stars that revolve around each other. It is estimated that 1/3 of the stars in our galaxy are binary or multiple. Two-body problems—we're talking about two masses, not the difficulty of finding jobs in nearby locations for a couple—have been mathematically solved.

The mathematics becomes much more difficult, however, when one considers three or more bodies. The classical situation involves the Moon–Earth–Sun system. The earth's coarse movement is dictated by the Sun, but the moon's mass and proximity make its effects felt on the Earth. Similarly, the Moon's movement is steered by the Earth, but the Sun exerts a nontrivial influence. A special case of a three-body problem occurs when one of the bodies is much lighter than the other two, for example, if we think of the Earth, the Moon, and a satellite. In this case, the mathematics simplifies significantly so that the equations may be solved. This is known as the restricted three-body problem.

In the 1770s, the French mathematician Joseph Lagrange found a fascinating result that occurs when all but two of the masses are relatively small. There are five points—the so-called *Lagrangian points* or *libration points*—where small bodies can remain stationary relative to the two larger bodies.

Figure 3.9 shows the five points in an idealized Sun–Earth configuration. Point $L1$ is an ideal location for making observations of the sun because it is never blocked by the Earth or the Moon. Several observatories—for example, the Solar and Heliospheric Observatory—are stationed in orbits near this point. Similarly, point $L2$ is ideal for space-based observations that seek to avoid the Sun's light; several observatories have been placed at this point as well. Point $L3$ is unstable because Venus comes relatively close every 20 months, and its gravitational pull would dislodge any light satellite. This hasn't stopped science fiction writers from postulating the existence of a "counter-Earth" at this point. The existence of points $L1$–$L3$ is not surprising because these three points, plus the two large bodies, are collinear. Points $L4$ and $L5$ are the real novelty found by Lagrange. Points $L3$–$L5$ lie on the orbit of the Earth and form an equilateral triangle.

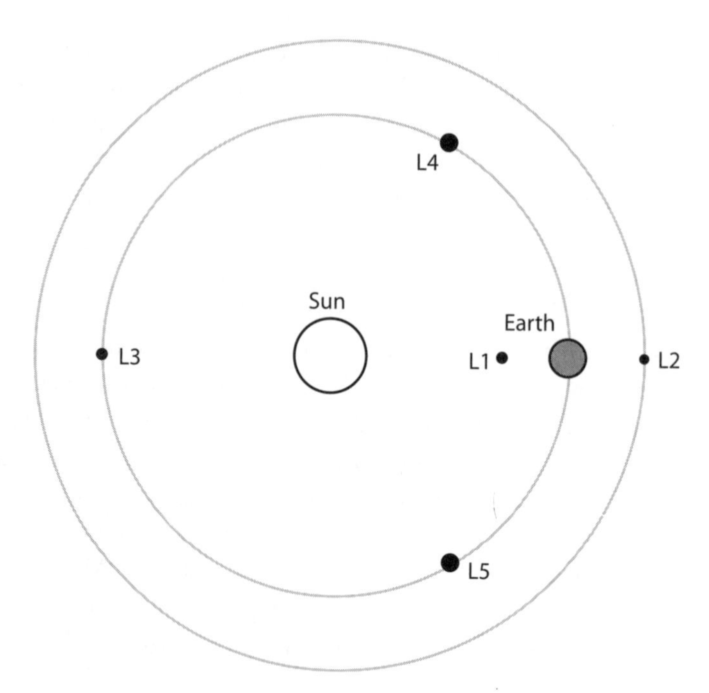

FIGURE 3.9: The libration points.

Point $L4$ is 60 degrees "ahead" of the Earth, while point $L5$ is 60 degrees "behind." In the Sun–Earth system, only interstellar dust can be found at points $L4$ and $L5$. In the Sun–Jupiter system, however, points $L4$ and $L5$ are the center of mass of the so-called Trojan asteroids. The first Trojan was spotted in 1906 by the German astronomer Max Wolf, and to date more than 4,000 Trojans have been observed at both points.

Notwithstanding the beautiful observations related to these special cases, the general three-body problem seems very difficult to crack. By the late 1800s, interest in the n-body problem had grown sufficiently strong that Gösta Mittag-Leffler, Sweden's most famous mathematician, advised King Oscar II to establish a prize for anyone who could solve this problem. The prize was eventually awarded to Henri Poincaré, who showed that an infinite series solution could not be used to solve the three-body problem. While numerical approximations to orbits can be constructed using Newton's equations over relatively short time frames—this is what is used in practice—Poincaré's work showed that an exact solution seemed illusory. This research was a harbinger for what would become chaos theory.

The Lorenz Attractor and Chaos

Atmospheric scientist Ed Lorenz got more than he bargained for when he numerically studied a problem in atmospheric convection. Physically, this phenomenon concerns the circulation of fluid in a shallow layer that is heated uniformly from below and cooled from above. The circulating fluid develops an apparently regular pattern called convection rolls. Lorenz studied a simplified mathematical model of this phenomenon that involved only three variables: the intensity of the convective motion, the temperature difference between the ascending and descending currents, and the distortion of vertical temperature profile from linearity, a positive value indicating that the strongest gradients occur near the boundaries. Mathematically, these three functions of time—x, y, and z—satisfy the differential equations

$$\frac{dx}{dt} = \sigma(y - x),$$
$$\frac{dy}{dt} = x(\tau - z) - y,$$
$$\frac{dz}{dt} = xy - \beta z$$

where σ, τ, and β are physically motivated constants. By specifying the values of x, y, and z at some initial time, this system of equations should uniquely determine the values of these variables for all future times.

When a system's future states are uniquely determined from its current state and the laws governing the evolution of that system, the system is called *deterministic*. A fundamental question in the area of dynamical systems asks for the long-term behavior of deterministic systems. What Lorenz found came as a complete surprise.

To give some context, let's think of a similar mathematical situation in two dimensions. Imagine tracking the orbits of points moving around the plane. Impose the condition that two distinct orbits cannot cross and that each orbit does not stray too far from its initial point. What is

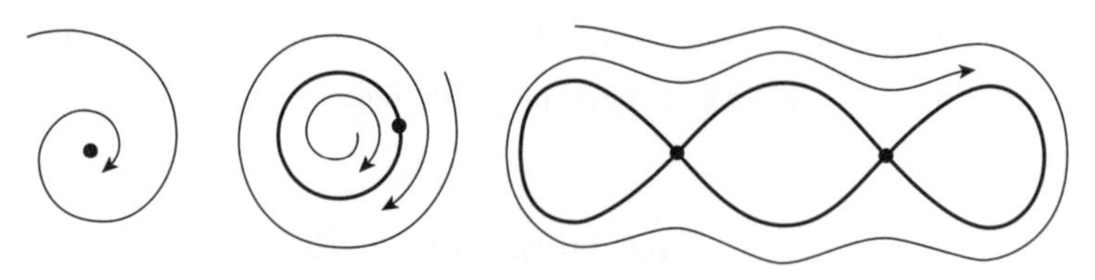

FIGURE 3.10: Approaching an equilibrium point, a limit cycle, or a cycle graph.

the possible long-term behavior of an orbit? Two obvious possibilities are that the point approaches an *equilibrium point* (a point that always stays fixed) or a *limit cycle* (a loop that repeats itself after a finite time). The Poincaré–Bendixson Theorem, proved in 1901, claims that there is only one other possible long-term behavior of an orbit: a *cycle graph*, which is a set of equilibrium points connected by orbits. All three limiting behaviors are depicted in figure 3.10. The Poincaré–Bendixson Theorem is used in applications ranging from mechanics to mathematical biology. This result, which is highly dependent on the noncrossing of orbits, guarantees highly structured behavior in two-dimensional systems.

Now back to the Lorenz equations. Suppose that the initial values of x, y, and z are specified and the orbits wander in *three*-dimensional space. What is the possible long-term behavior of the orbits? Are we restricted to the three-dimensional analogues of the options laid out in the Poincaré–Bendixson Theorem? Lorenz stumbled upon a very bizarre limiting set. This complicated set—depicted in figure 3.11—is now called the *Lorenz attractor*. It may be mistaken as a cycle, but it's much more than that. Any nearby point gets sucked into the Lorenz attractor. The attractor itself contains both order and disorder in that there are infinitely many cycles, but they are all repelling. This strange set also contains dense orbits that never repeat but come arbitrarily close to any other point in the attractor.

To distinguish the Lorenz attractor from the banal possibilities enumerated by the Poincaré–Bendixson Theorem, this set is an example of what is called a *strange attractor*. This exotic set is a prototype of chaos in continuous systems. Far from being rare, strange attractors have appeared in models in physics, chemistry, biology, electronics, and economics.

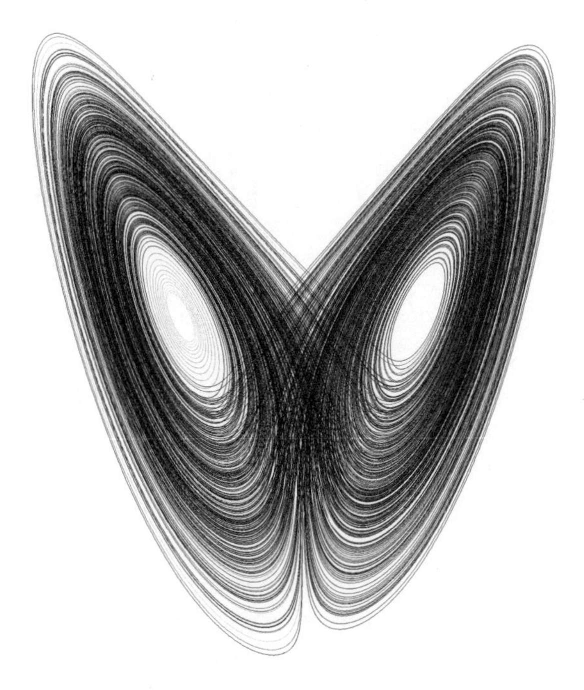

FIGURE 3.11: The Lorenz attractor.

Period Three Implies Chaos

Physics has long been the main beneficiary of mathematical help. Mechanics, electromagnetism, and relativity all boast theories beautifully expressed and explained in the language of mathematics. The last century saw cautious advances of mathematics in other fields such as economics and biology. Things get slippery there because phenomena in these newer fields are apparently not restricted by tidy "laws" such as Newton's equations of motion or Maxwell's equations for electromagnetism. And yet, there is sometimes order observed on large scales, order that begs for attention.

Consider, for example, human population growth in the twentieth century. While many factors affect birth and death rates—health habits, spread of diseases, wars, proliferation of weapons, availability and use of birth control, directives of religious and government entities, to name a few—observations correlate roughly with Malthusian growth over short time frames. This principle asserts that the rate of growth at a

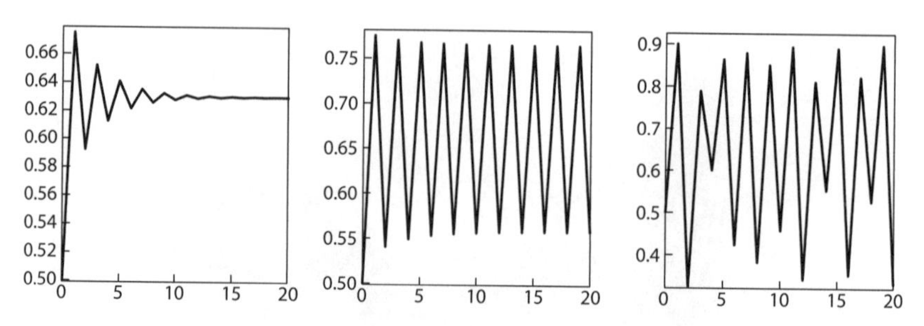

FIGURE 3.12: Dynamics for $r = 2.7$ (left), 3.1 (center), and 3.6 (right).

given time is proportional to population level at that time. Letting p_n represent the population in year n, the Malthusian model is

$$p_{n+1} = rp_n \tag{3.1}$$

for some constant $r > 1$. The constant r can be interpreted as the ratio of the populations in two successive years. Like compound interest, however, this exponential growth in the population size accelerates each year, a state clearly unsustainable because of the limitation on resources (food, air, land, take your pick). To correct this long-term problem in the model, mathematicians add an extra term to equation (3.1):

$$p_{n+1} = rp_n - sp_n^2 \tag{3.2}$$

for some constant $s > 0$. If p_n is too large, then $p_{n+1} < p_n$, so this new equation possesses built-in self-limitation on the population size. Equation (3.2) involves two parameters (r and s) and can be transformed into a simpler one-parameter factored equation:

$$x_{n+1} = rx_n(1 - x_n) \tag{3.3}$$

where r is now any positive constant.

Now the mathematicians elbow the biologists out of the way to ask about the long-term dynamics of equation (3.3). What happens for different choices of the parameter r? Figure 3.12 shows the dynamics for three choices of r, in each case starting at $x_0 = 0.5$. If $0 < r < 1$, the population shrinks each year and approaches extinction.

If $1 < r < 3$, the population persists and approaches a constant level. When r is slightly larger than 3, another qualitative change occurs and the population approaches a 2-cycle: it alternates between two different values. It's important to state that the constant function approached in the long term when $1 < r < 3$ has not disappeared; it is now simply repelling. Most initial states x_0 will approach the 2-cycle, but special choices of x_0 will approach the constant function. This shift in qualitative behavior witnessed as r increases across a threshold value is called a *bifurcation* (this special kind of bifurcation is actually called a *period-doubling bifurcation*). As the parameter r increases further, the 2-cycle bifurcates into a 4-cycle, then to an 8-cycle, etc. Again, the other cycles have not disappeared; they simply turned from being attracting to repelling. When $r \approx 3.57$, the period-doubling climaxes and we enter the *chaotic region*. Both the graphics and the mathematics become much more complicated. Finding order in this new region is far from obvious. It turns out that for each r value where $0 < r < 4$, there is exactly one attractive cycle and all the other cycles are repelling.

The mathematical world was caught by surprise with a 1975 paper by Tien-Yien Li and James Yorke titled, "Period Three Implies Chaos." They showed that when there is a cycle of length 3 (attracting or repelling), there is a cycle of *every* other period! A 3-cycle is a marker of very complicated behavior. Despite the crazy dynamics, windows of sanity can be detected. For example, if $r \approx 3.83$, there is an attractive 3-cycle, showing aplomb amid the surrounding chaos.

Patterns among the Stars

Mathematics could be generally described as an effort to systematize patterns. The human need for order—except for teenagers' bedrooms—predates civilization and is evidenced in the perceived patterns found among the stars. Finding order in large, random data sets has been studied by mathematicians through Ramsey Theory. Generally speaking, this area studies the question, "How many objects of type X need to be present to witness pattern Y?"

An example involves a beautiful result concerning the number 3. Birch's Theorem claims that from any set of $3N$ points in the plane

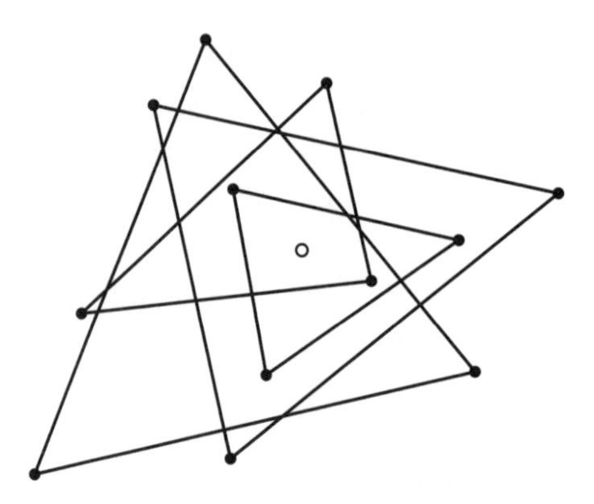

FIGURE 3.13: From 12 solid points, 4 triangles are made that share a common hollow point.

one can construct N triangles that share a common interior point. Figure 3.13 gives an example with $N = 4$.

Fermat's Last Theorem

The twentieth century closed with a proof of Fermat's Last Theorem, a chestnut of number theory that had been stated more than 350 years earlier. Its proof, accomplished by Andrew Wiles with the help of his former student, Richard Taylor, was heralded in the popular press around the globe. Mathematics is not typically front-page news, but then, everyone enjoys the resolution of a good mystery.

Let's motivate the theorem with a simple question. Which positive integers a, b, and c satisfy the equation $a^2 + b^2 = c^2$? This is an example of a Diophantine equation, named after the ancient Greek mathematician Diophantus of Alexandria, who studied many similar equations. For this particular equation, all the solutions may be written explicitly:

$$a = m^2 - n^2, \quad b = 2mn, \quad c = m^2 + n^2$$

where m and n are any integers and a and b can be interchanged. For example, setting $m = 5$ and $n = 2$ produces $a = 21$, $b = 20$, and $c = 29$. In 1637, the French mathematician Pierre de Fermat wondered about solving similar Diophantine equations, specifically, the same equation but with higher powers:

$$a^n + b^n = c^n \tag{3.4}$$

where $n \geq 3$ is an integer. Fermat claimed that there were no positive integers a, b, and c satisfying equation (3.4). In his copy of Diophantus' *Arithmetica*, Fermat had scribbled several of his own mathematical claims in the margins but omitted detailed proofs. Later mathematicians examined Fermat's assertions and disposed of all the unsubstantiated claims—all but one, the question about the higher powers. Euler, Gauss, Dirichlet, and many giants of mathematics cut their teeth on Fermat's claim but couldn't get far. It became known as Fermat's Last Theorem.

Like many simply stated mathematical problems that stubbornly resist defeat, Fermat's Last Theorem has suffered its share of fools, professional and otherwise. To encourage efforts on this problem, the Wolfskehl Prize of 100,000 German marks was established in 1908 for a complete proof. Many questionable people waltzed along in the hopes of snagging this prize money; over the next four years, reportedly more than 1,000 false proofs were published. While most of these were amateurish attempts, some professional mathematicians have shuffled away in shame when their published "proof" was found to have an error. Even giants like Euler and Lamé had false starts (but both ended up making solid, albeit small, contributions). Conquering Fermat's Last Theorem had become something like the Triple Crown of mathematics.

A pivotal moment in the history of the problem came in 1984 when Gerhard Frey claimed that Fermat's Last Theorem was true if one could prove the Taniyama–Shimura Conjecture, a fantastical conjecture made in the 1950s. In 1986, Frey's connection was formally proved, so now settling Fermat's Last Theorem could be accomplished if the Taniyama–Shimura Conjecture could be solved. The problem was that no one had any idea how to prove this conjecture either. Enter Andrew Wiles.

Upon learning that Frey's connection was true, Wiles, a British mathematician who had become a professor at Princeton University, dropped other research projects and devoted himself entirely to proving

Taniyama–Shimura. This problem would not be resolved with "I came, I saw, I conquered" efficiency. For seven years, Wiles painstakingly worked on this problem in absolute secrecy. There are two obvious reasons for this guarded approach. First, embracing long-unsolved problems typically earns one the label of a "crank," an unstable person with unorthodox ideas. Second, he did not want his ideas to get snatched by others. With few exceptions, mathematical formulas do not have ownership. Companies like IBM or government organizations like the National Security Agency employ mathematicians and protect their findings, but generally speaking, mathematical discoveries are published openly. Wiles wanted to press through to a final solution instead of publishing his incremental progress.

After years of closeted work, Wiles unveiled his proof in 1993 to a startled world. During the vetting process, however, an apparently insurmountable mistake was found. With the help of Richard Taylor, Wiles clearly navigated around the error and the corrected proof was accepted in 1994. The papers published by Wiles and Taylor explaining the proof are more than 100 pages of dense, advanced mathematics that only a handful of experts can understand.

While the proof was heralded as a monumental work of a great mathematical mind and was trumpeted far and wide, some mathematicians have asked, "Is there not an easier proof?" None has been found, but an interesting 100-year-old result touches on the $n = 3$ case. Although Euler essentially settled this case much earlier, an interesting theorem due to Ramanujan brings further light to this problem. Define three sequences $\{a_n\}$, $\{b_n\}$, and $\{c_n\}$ implicitly by the following generating functions:

$$\sum_{n=0}^{\infty} a_n x^n = \frac{1 + 53x + 9x^2}{1 - 82x - 82x^2 + x^3}$$

$$\sum_{n=0}^{\infty} b_n x^n = \frac{2 - 26x - 12x^2}{1 - 82x - 82x^2 + x^3}$$

$$\sum_{n=0}^{\infty} c_n x^n = \frac{2 + 8x - 10x^2}{1 - 82x - 82x^2 + x^3}$$

Table 3.1
Ramanujan's Sequences

n	a_n	b_n	c_n
0	1	2	2
1	135	138	172
2	11,161	11,468	14,258
3	926,271	951,690	1,183,258
4	76,869,289	78,978,818	98,196,140
5	6,379,224,759	6,554,290,188	8,149,096,378

The first few values are given in table 3.1. Ramanujan came up with a beautiful formula relating the three sequences:

$$a_n^3 + b_n^3 = c_n^3 + (-1)^n$$

for $n = 0, 1, 2, \ldots$. This produces infinitely many "near misses" to the Fermat equation. It also makes one wonder about near misses for the Fermat equation with higher values of n.

Leftovers Anyone?

A refrigerator bulging with leftovers is not anyone's idea of a happy sight (or smell). But leftovers in math, namely remainders from a division process, can offer delicious ideas.

An ancient result buried in the back of the fridge—and used frequently by number theorists—is called Fermat's Little Theorem (don't confuse this with Fermat's Last Theorem). This claims that if p is a prime and a is an integer that is not a multiple of p, then

$$a^{p-1} \equiv 1 \pmod{p} \qquad (3.5)$$

which means that a^{p-1} divided by p has a remainder of 1. Mathematicians call this a congruence relation. An important application of this theorem is in primality testing. Given a number p, it can be shown that it is *not* prime if there is a number a where the relationship (3.5) fails. For example, $2^8 \equiv 4 \pmod{9}$, hence 9 is not a prime number. If p passes this test for many values of a, p is called a *pseudoprime*. However, there are composite numbers p that satisfy relationship (3.5) for *all* a. Such values of p are called *Carmichael numbers*. The smallest Carmichael number is 561, and in 1994 it was shown that there exist infinitely many Carmichael numbers.

Another well-known congruence relation is Wilson's Theorem: a number p is prime if and only if $(p-1)! \equiv -1 \pmod{p}$. While this result characterizes primes, it is useless as a primality test because of the demands in computing $(p-1)!$. Besides looking at congruence relations for powers (Fermat's Little Theorem) or factorials (Wilson's Theorem), some results have been found for binomial coefficients. The classical result here is called Lucas's Theorem: if p is a prime, $0 \leq n$ and $j < p$, then

$$\binom{pm+n}{pi+j} \equiv \binom{m}{i}\binom{n}{j} \pmod{p}$$

All of these congruences involve division by p. However, there are much stronger results with division by p^3. One is Morley's Congruence: If $p > 3$ is prime, then

$$(-1)^{(p-1)/2}\binom{p-1}{(p-1)/2} \equiv 4^{p-1} \pmod{p^3}$$

The next result, known as Wolstenholme's Theorem, is similar: If $p > 3$ is prime, then

$$\binom{2p-1}{p-1} \equiv 1 \pmod{p^3}$$

There is no known composite value of p that satisfies this congruence, leaving one to wonder if this relationship also characterizes primes. And lest one get greedy, this congruence relationship does not extend to fourth powers. Aren't cubes of primes spicy enough?

Egyptian Fractions

According to the ancient Egyptians, fractions whose numerators are 1—called *unit fractions*—are "purer" than other fractions; those with numerators greater than 1 were sometimes referred to as *vulgar fractions*. Of course, every vulgar fraction m/n can be written as a sum of unit fractions: Just add m terms to get $1/n + \cdots + 1/n = m/n$. The Egyptians asked for the additional constraint that the denominators of each unit fraction be distinct. We call such a form an *Egyptian fraction*. The Rhind Mathematical Papyrus, dating back to about 1650 BCE, contains a table of Egyptian fractions for fractions of the form $2/n$.

Every fraction can be written as an Egyptian fraction. One way to do this is to recursively use the formula

$$\frac{1}{k} = \frac{1}{k+1} + \frac{1}{k(k+1)}$$

For example,

$$\frac{2}{7} = \frac{1}{7} + \frac{1}{7} = \frac{1}{7} + \frac{1}{8} + \frac{1}{56}$$

and

$$\begin{aligned}\frac{3}{7} &= \frac{2}{7} + \frac{1}{7} \\ &= \left(\frac{1}{7} + \frac{1}{8} + \frac{1}{56}\right) + \left(\frac{1}{8} + \frac{1}{56}\right) \\ &= \frac{1}{7} + \frac{1}{8} + \frac{1}{56} + \frac{1}{9} + \frac{1}{72} + \frac{1}{57} + \frac{1}{56 \cdot 57}\end{aligned}$$

Another approach is to subtract the largest possible unit fraction from the starting fraction and apply this approach iteratively to the remainder until one is left with a unit fraction. For example,

$$\begin{aligned}\frac{4}{625} &= \frac{1}{157} + \frac{3}{98,125} \\ &= \frac{1}{157} + \frac{1}{32,709} + \frac{2}{3,209,570,625} \\ &= \frac{1}{157} + \frac{1}{32,709} + \frac{1}{1,604,785,313} \\ &\quad + \frac{1}{5,150,671,800,036,230,625}\end{aligned}$$

Note that the numerator of the remainder decreases with each step. This proves that m/n can be written as an Egyptian fraction with at most m terms. In mathematics in general, procedures like this, which hoard as much as possible in each step, are called *greedy algorithms*. This greediness allows one to reach his or her goal usually with the fewest number of steps. While the procedure is easy to use, this process may not produce the fewest number of terms. Although every number of the form $4/n$ can be written as the sum of four distinct unit fractions, the last example may be written in many ways with only *three* unit fractions:

$$\frac{4}{625} = \frac{1}{160} + \frac{1}{6,667} + \frac{1}{133,340,000} = \frac{1}{200} + \frac{1}{715} + \frac{1}{715,000}$$
$$= \frac{1}{240} + \frac{1}{448} + \frac{1}{840,000} = \frac{1}{250} + \frac{1}{417} + \frac{1}{521,250}$$
$$= \frac{1}{375} + \frac{1}{268} + \frac{1}{502,500} = \frac{1}{450} + \frac{1}{240} + \frac{1}{90,000}$$
$$= \frac{1}{500} + \frac{1}{228} + \frac{1}{71,250} = \frac{1}{750} + \frac{1}{198} + \frac{1}{61,875}$$

The Erdős–Straus Conjecture claims that every fraction of the form $4/n$ is the sum of three distinct unit fractions. This problem was posed in 1948 and is still unresolved.

In the previous example, a choice with smaller denominators is

$$\frac{4}{625} = \frac{1}{250} + \frac{1}{500} + \frac{1}{2,500}$$

This was found by using $4/5 = 1/2 + 1/4 + 1/20$ and dividing each term by 125. This example highlights that to prove the Erdős–Straus Conjecture, it is sufficient to consider only the fractions $4/n$ where n is a prime. In an effort to show that large swaths of numbers fulfill the Erdős–Straus Conjecture, some have developed formulas that explicitly construct the three unit fractions when n has a certain structure.

Table 3.2
Voting Preferences for a Class of 20 Students

Preference	*Number*
B > D > V	4
B > V > D	3
D > B > V	5
D > V > B	0
V > B > D	2
V > D > B	6

For example, if $n \equiv 2 \pmod 3$, then

$$\frac{4}{n} = \frac{1}{n} + \frac{1}{1 + (n-2)/3} + \frac{1}{n(1 + (n-2)/3)}$$

In fact, more work like this shows that any counterexample to the conjecture must satisfy $n \equiv 1 \pmod{24}$.

Arrow's Impossibility Theorem

Compared to the challenges of chaos theory, the three-body problem, and atmospheric dynamics, one would imagine that mathematical questions about voting would be easy to answer. After all, two candidates are running for a position and the one who receives the most votes wins. Simple, right? But what if three candidates are in the mix? Things can get complicated in a hurry. For example, suppose it's a rainy day and the gym class is offered the choice of three indoor activities: basketball (B), dodgeball (D), or volleyball (V). When the students' preferences for each activity are polled, the resulting cacophony propels the teacher to give his or her students a lesson in the subtlety of voting.

He or she first passes out slips of paper and asks each student to record his or her preferences in order. For example, if a student's preference is basketball, dodgeball, and lastly volleyball, he or she should write B>D>V. The class's preferences are indicated in table 3.2.

In a typical election, only the voter's first choice is valued. This is called *plurality* voting, and with this approach, volleyball is the winner: V:8, B:7, D:5. Before you put on your knee pads, however, let's take a more nuanced look at the voting data. We have a lot more information than simply each voter's first choice. When the teacher initially polled the class on its preferences, some students howled at the mention of some sports. They seemed to suggest that two activities were fine, but the other was a distinct third. Suppose that instead of counting only the first vote, we assigned one point for each of the first two preferences. This strategy, known as *antiplurality* voting, is equivalent to voting against the voter's last choice. This approach produces D: 15, B: 14, V: 11, so dodgeball wins.

The gym teacher, however, is not done. He or she reasons that the antiplurality vote does not seem so fair because it equally values the voter's first and second choices. A more balanced approach gives two points to a voter's first choice, one point to his or her second choice, and zero points to the third choice. This strategy, known as the *Borda count*, yields B: 21, D: 20, V: 19, making basketball the winner.

What have we seen? With three or more "candidates" and each voter's preferences, different ways of assessing votes produce different results. Lest one think that such matters are abstract mathematical examples, consider the 2000 presidential election in the United States, the infamous challenge between George W. Bush and Al Gore. And don't forget Ralph Nader. Every federal election has candidates from many political parties besides Republicans and Democrats—the Communist Party, the Christian Liberty Party, the Marijuana Party, etc.—but in 2000, the Green Party's Ralph Nader garnered 2.7% of the popular vote, enough to make a difference. Popular wisdom suggests that Nader's supporters largely came at the Democrats' expense, so if plurality voting had not been used, Al Gore might have won the election.

The waters for voting are actually muddier than you think. Long before the hanging chads of the 2000 election, efforts to find fair voting systems had been made. A surprising result known as Arrow's Impossibility Theorem helped win Kenneth Arrow the Nobel Prize in economics in 1972. The theorem starts by asking that a voting system satisfy a few reasonable assumptions:

No dictators: The outcome is not simply the wish of one person.

Pareto efficiency: If every voter prefers candidate A to candidate B, then the outcome should have A over B.

Independence of irrelevant alternatives: If voters change their preferences between candidates other than A and B, this should not affect the outcome between A and B.

Arrow's theorem claims that no election method with three or more candidates can satisfy all of these conditions. This does not mean the end of democracy, but it does suggest that care should be taken in the democratic process. Researchers in this area explore questions such as whether some systems are fairer than others.

Mapping Surfaces

The coarse shape of our planet is more accurately measured with an ellipsoid as opposed to a sphere. Why? While gravity's effects mold a large enough mass into a sphere over time (to minimize potential energy), the spin of the planet fattens it up around its equator. Large scale aside, it's clear that these global approximations do not capture the subtlety of towering mountains or ocean trenches. Modeling the surface of the Earth—and other surfaces such as a person (figure 3.14)—requires more nuance.

To approximate a surface, first find a large set of its points. Next, use these points to construct a *triangulation*. A triangulation connects the set of points with line segments so that the surface is approximated by a collection of triangles. Every surface has a triangulation, and there are several approaches to generate one. For the purposes of producing an attractive mesh, a well-used method is Delaunay triangulation. This technique maximizes the minimum angle in any triangle, thus helping to avoid skinny triangles. While the triangulation can accurately approximate the surface, its rendering suffers from having sharp edges and corners. To remedy this issue, one can approximate each triangular face with a function that joins smoothly to the functions on the neighboring triangles. These smoothing functions, called splines, are

FIGURE 3.14: Triangulation of a person.

often cubic polynomials. A huge amount of research has gone into each of these steps for mapping a surface. The success of these methods is evident in that the end result completely disguises this effort.

Guarding an Art Gallery

With budget cuts, a city's art gallery has to economize on the number of guards it employs. Assuming that the guards on duty are stationary and that every part of the gallery must be in at least one guard's line of

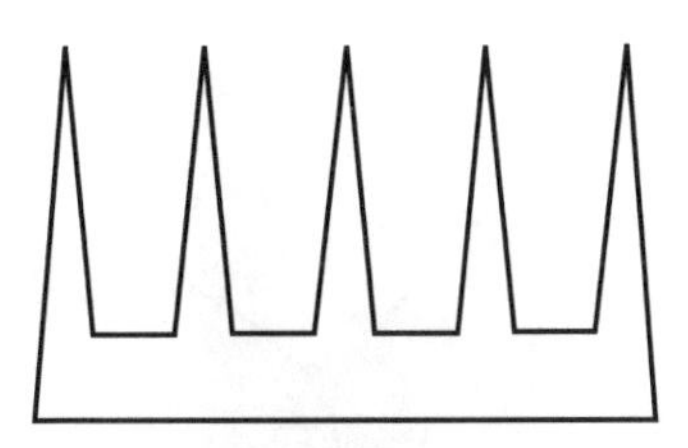

FIGURE 3.15: This gallery with 15 walls needs 5 guards.

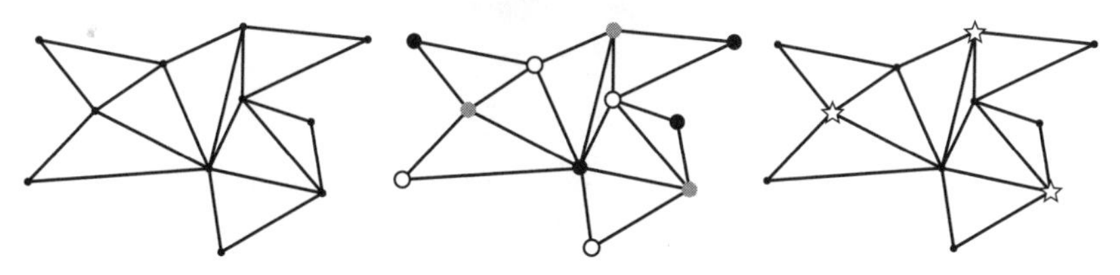

FIGURE 3.16: Triangulation (left), coloring (center), and choosing the guard positions (right).

sight, what is the minimum number of guards necessary to completely protect the gallery? Of course, this depends on the shape of the gallery. If the gallery was simply a rectangular room, clearly one guard would be sufficient. What if the gallery was a polygon with n sides?

It's easy to produce examples where $n = 3k$ and k guards are needed. Construct k thin culs-de-sac branching off from a long corridor; figure 3.15 gives an example with $k = 5$. This configuration has been called *Chvatal's comb*.

Remarkably, the Art Gallery Theorem claims that this is the worst-case scenario. Specifically, for a gallery with n sides, no more than $\lfloor n/3 \rfloor$ guards are needed. A proof due to Steve Fisk in 1978 is deceptively simple and beautiful. It starts with triangulating the region, that is, adding extra edges between the vertices of the polygon so that each subregion is a triangle (figure 3.16). The next step involves coloring each vertex with one of three colors so that each triangle uses each color exactly once. Such a coloring is always possible. Lastly, identify the color that is used the least and place a guard at each of those vertices. Every triangle—and hence the whole gallery—is now guarded.

The Poincaré Conjecture

Imagine an ant crawling on the surface of an oil drum. Although the top is flat and the sidewall is cylindrical, the ant only senses nearby

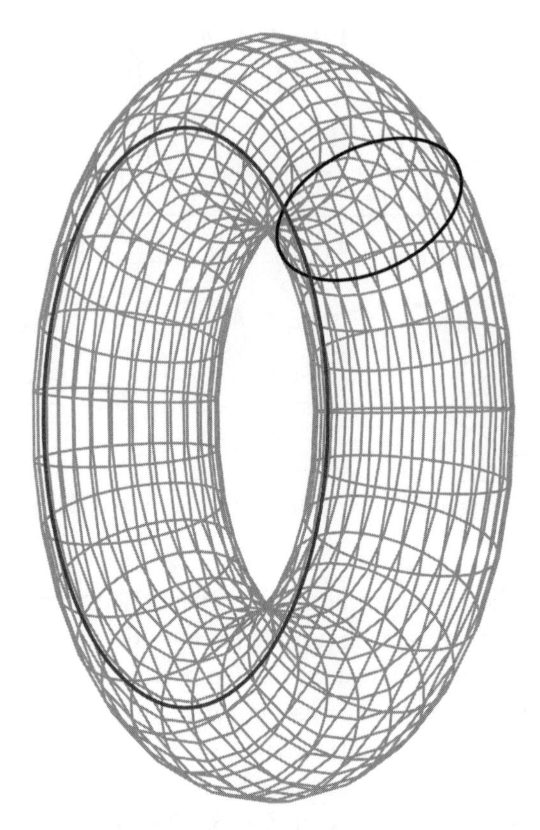

FIGURE 3.17: Unshrinkable loops on a torus.

points on this large surface and so either portion of the drum appears flat. The "bending"—mathematicians say *curvature*—of the surfaces is irrelevant; the ant simply detects a smooth surface. Don't laugh at the poor ant, however; only recently in our history have we realized that the Earth is not flat. Surfaces that look flat nearby are formally called two-dimensional manifolds, or 2-manifolds.

Let's restrict our attention to only special 2-manifolds. First, assume that each surface is *closed*, that is, has no edge. A disk floating in space has an edge, but an ellipsoid does not. Second, we want the surface to be *simply connected*. One way to describe this is that any closed loop on the surface can be continuously deformed—while staying on the surface—and shrunk to a point. This works for a sphere, but not for a donut (figure 3.17); neither of the marked curves can be shrunk to a point.

Now we can state the main warm-up theorem: any simply connected, closed 2-manifold is a sphere. The word "is" means we can deform the surface—stretching is okay, tearing is not—into a standard sphere. In other words, any surface that has no edge and allows every loop to shrink to a point is just a stretched or distorted version of a sphere. In fact, there is a theorem that classifies all closed 2-manifolds, but we want to stay on topic. The general result—the classification of closed 2-manifolds—dates back to the 1860s.

At the dawn of the twentieth century, Poincaré was thinking about a harder problem. He was interested in higher dimensional surfaces known as 3-manifolds. This is hard to picture, but just as the ant senses that each part of the surface is locally two-dimensional, a 3-manifold seems locally three-dimensional. Just as the 2-sphere can be described algebraically as $x^2 + y^2 + z^2 = 1$, the 3-sphere $x^2 + y^2 + z^2 + w^2 = 1$—we're representing four-dimensional space with x, y, z, and w axes—is the simplest example of a 3-manifold. Poincaré's question asks whether every simply connected, closed 3-manifold is a 3-sphere. He couldn't answer this question.

Sometimes when mathematicians can't answer the question at hand, they look at a more general problem in the hope of gaining more insight. This strategy led to the generalized Poincaré Conjecture, which asks the same question for n-manifolds where $n \geq 3$. In 1961, Stephen Smale proved this conjecture when $n \geq 5$ and Michael Freedman proved the $n = 4$ case in 1982. Only Poincaré's original case, $n = 3$, was left standing. In 2002 and 2003, the Russian mathematician Grigori "Grisha" Perelman posted three articles on the Internet that proved the conjecture. The journal *Science* listed the proof of the Poincaré Conjecture as the "breakthrough of the year" in 2006.

Perelman's solution and the ensuing drama have attracted attention from far and wide. First, his results were not published in a peer-reviewed journal but posted on a website that simply shares scientific papers as received. Without validation from experts, there was no recognized stamp of approval on his work. Second, Perelman was offered a Fields Medal—something like a Nobel Prize in mathematics—in 2006 for his work, but he turned it down. His scruples for turning down this prestigious award raised many eyebrows. In essence, Perelman claimed that he deserved this honor no more than the others who had built the mathematical machinery used in his proof. But another twist was yet to come. The Poincaré Conjecture is one of the seven Millennium Prize Problems listed by the Clay Mathematics Institute. In 2010, watchers were not surprised to learn that Perelman turned down this honor (and money) as well.

Monge's Three-Circle Theorem

Plenty of geometry theorems involve three shapes, or three in a row, etc. One of these is Monge's Theorem: Given three disjoint circles in the plane with distinct radii, the three intersections points of pairwise tangent lines are collinear (figure 3.18).

This theorem is noteworthy because it is a *two*-dimensional result that has a lovely *three*-dimensional proof. Sometimes getting a bird's-eye view gives the needed perspective. Replace each circle with a sphere whose radius and center are the same. One can now picture the original plane as a water's surface and the three spheres as half-submerged. For each pair of spheres, construct the unique cone that is tangent to the two spheres. The apex of the cone must lie on the line passing through the centers of both spheres, so it must lie on the plane as well. Now here's the really clever idea. Imagine another plane resting on top of the spheres. This new plane must pass through the apex of each cone. Since this new plane intersects the first plane in a line, the three points of interest must be collinear.

Marden's Theorem

When two people meet and one of them says, "I'm a mathematician," the responses are somewhat predictable. After an awkward pause, the other person often exclaims, "I was never very good at math." The mathematician wonders why when the same person meets a physician they don't exclaim, "I was never very good at biology." In any case, another common response to the mathematician is "Isn't all of mathematics known?" Thousands of mathematical journal articles with new theorems are written each year, although admittedly they are at a very high, technical level, which makes explaining them to the average person nearly impossible. Every once in while, though, a new result is found that is both beautiful and relatively easy to explain. Such is the

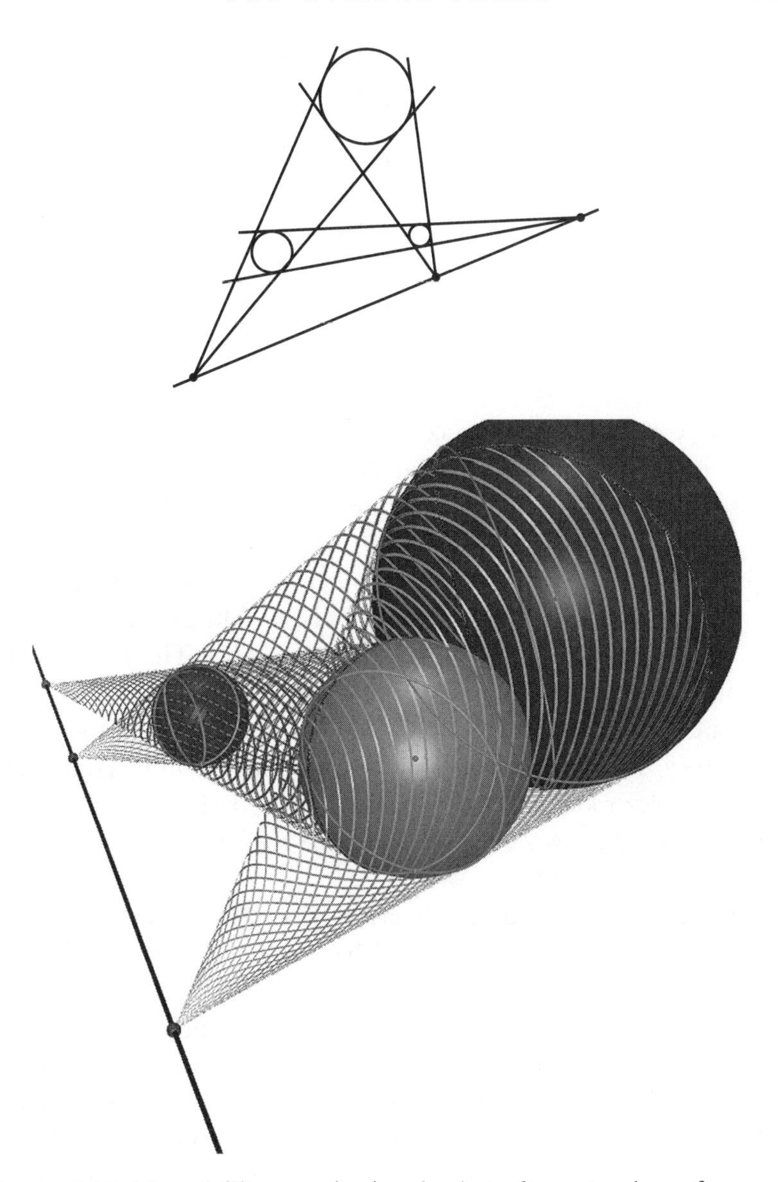

FIGURE 3.18: Monge's Theorem (top) and a three-dimensional proof (bottom).

case with the gem called Marden's Theorem. Morris Marden mentions the result in a 1945 paper but gives credit to Jörg Siebeck for finding it almost 81 years earlier.

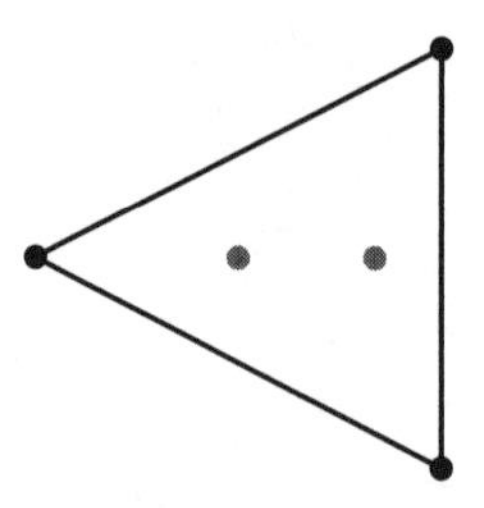

FIGURE 3.19: Roots of f surrounding the roots of f'.

Some context will be instructive before we state Marden's Theorem. What is the relationship between the zeros of a polynomial and the zeros of its derivative? The function $f(x) = (x-1)(x-2)(x-3)$ has zeros at $x = 1, 2, 3$, while $f'(x) = 0$ when $x = 2 \pm 1/\sqrt{3}$, approximately 2.58 and 1.42. Note that these two roots of f' are between 1 and 3, two of the roots of f. Now shift the function up by 10 units so that $f(x) = (x-1)(x-2)(x-3) + 10$. This new function has only one real root, $x \approx -0.31$, but since f' has not changed, its roots are the same, so the roots of f' do not lie between the roots of f. Or do they? Let's consider all the roots of f, not just its real roots. The three complex roots of f are approximately -0.31 and $3.15 \pm 1.73i$. The roots of f', $x = 2 \pm 1/\sqrt{3}$, lie inside the triangle formed by the roots of f (figure 3.19).

This is not a coincidence. Every degree three polynomial f has exactly three complex roots, and f' has two roots. The two roots of f' always lie within the triangle whose vertices are the roots of f. This is a special case of the Gauss–Lucas Theorem, which asserts that for any polynomial f, the roots of f' lie within the convex hull of the roots of f. What is the convex hull of a set? A set is *convex* if the line segment joining any two points in the set also lies completely in the set. The *convex hull* of a set S is the smallest convex set that contains S. The convex hull can be pictured by imagining a rubber band squeezing tightly around the points in the set. This set of surrounded points constitutes the convex hull of S.

So with all of this background, what is Marden's Theorem? Let's go back to the case of a polynomial f of degree three. We know that the roots of f' lie within the triangle whose vertices are the roots of f

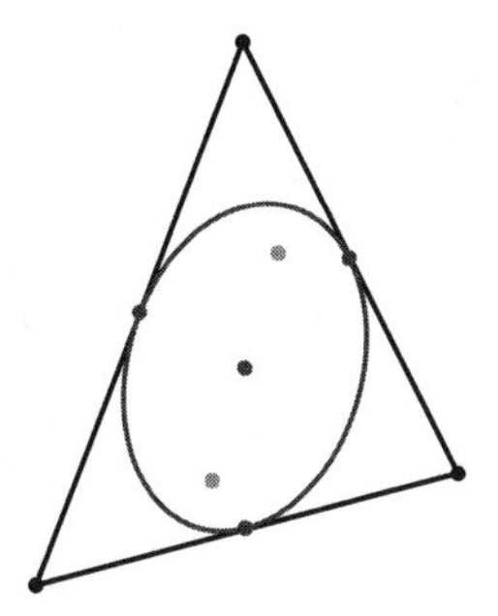

FIGURE 3.20: Marden's Theorem.

(we're avoiding the degenerate case when the roots are collinear). Generate the unique ellipse that lies within the triangle and is tangent to the midpoints of the three sides (figure 3.20). Marden's Theorem claims that the two roots of f' are the foci of the ellipse! This wholly unexpected result does not require advanced mathematics to prove, but neither is it something so simple that an expert could dispose of its proof in a moment. A 2008 journal article about Marden's Theorem was called "The Most Marvelous Theorem in Mathematics." While most mathematicians would disagree with such a strong sentiment, there is no doubting that this is a beautiful result. That this simply stated theorem was discovered centuries after roots had been intensely studied is remarkable. Oh, and as a bonus, the single root of f'' is at the center of the ellipse.

The Reuleaux Triangle

Why does a manhole cover have a circular cross section? Disks have been ubiquitous since the invention of the wheel, but that's not the reason for the cover's shape. This lid could have a square cross section, a seemingly easier shape to construct. The problem is that once this heavy cover—usually more than 50 kg or 110 lb—is removed, it could easily drop at an angle and fall into the hole. To avoid this problem, the hole's cross section should have the same maximum width in every direction. Of course, the circle has this property since each maximum width is simply the diameter of the circle. Any shape that has the same width

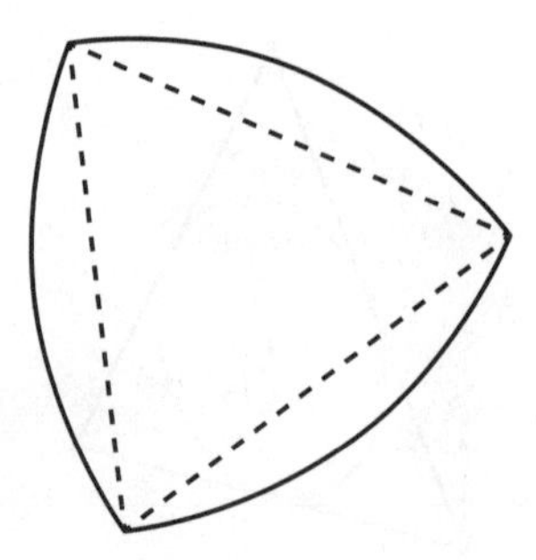

FIGURE 3.21: The Reuleaux triangle.

in each direction is called a *curve of constant width*. The simplest noncircular curve of constant width is the Reuleaux triangle (figure 3.21). Named after Franz Reuleaux, a German engineer, this curve is made up of three circular arcs. Each "corner" of the Reuleaux triangle is the center of the circle that generates the opposing arc. Barbier's Theorem states that the perimeter of a curve of constant width equals the width times π. The area does not have a similar property. The Blaschke–Lebesgue Theorem asserts that the Reuleaux triangle has the smallest area of all curves of constant width with the same radius.

A mechanical property of the Reuleaux triangle is that it can be used to drill nearly square holes. The drill bit whose cross section is a Reuleaux triangle won't work in a standard drill, though; to get the square hole, the axis of rotation must move in a circle while the bit is spinning. Special drill chucks have been designed to accomplish this feat. Speaking of mechanical connections, the Reuleaux triangle is sometimes confused for the rotor used in the Wankel engine.

In popular culture, the Reuleaux triangle has been used in signage and logos, but curves of constant width are best recognized as the shapes of certain coins. In Britain, the 20 pence and 50 pence coins are curves of constant width with seven "sides," a design used for easier identification in vending machines. The Canadian dollar coin—typically called a "Loonie" because of the common loon featured on one side—and the U.S. Susan B. Anthony dollar coin are curves of constant width that have 11 "sides." Three of these coins are shown in figure 3.22.

FIGURE 3.22: Coins of constant width: the British 50 pence (top), the Canadian dollar (middle), and the U.S. Susan B. Anthony dollar (bottom).

Curves of constant width can be smooth or can have an arbitrarily large numbers of corners. If such a curve has corners, like the Reuleaux triangle, a smoother version can be made by rolling a circle around the curve and tracing the outside of the path. Usually a curve of constant width does not have a compact, analytic representation, but the equation

$$(x^2+y^2)^4-45(x^2+y^2)^3-41,283(x^2+y^2)^2+7,950,960(x^2+y^2)$$
$$+16(x^2-3y^2)^3+48(x^2+y^2)(x^2-3y^2)^2$$
$$+(x^2-3y^2)x[16(x^2+y^2)^2-5,544(x^2+y^2)+266,382]=720^3$$

represents a curve of constant width.

Getting back to manhole covers, the circular hole still makes sense over other curves of constant width because it is easier to construct and requires no rotation for placement. The question of why manhole covers are circular was apparently popular in Microsoft job interviews.

The Third Critical Point

For those whose nerves are strained by roller coasters, perhaps the toughest part in the ride is when they just go over the first peak; the plunge is imminent. Of course, between any two peaks is a nadir, a low point. This is the case with any continuous function of one variable. If the function has two local maxima—points where the function is at least as high as any close point—then there must be a local minimum between them.

A harder question is the analogue for functions of two variables. If a continuous function has two local maxima—imagine these as hilltops, not just the silhouettes from the roller coaster—must there be a local minimum between them? This is actually not the right question to ask. For functions of two variables, maxima and minima are not the only typical critical points. Calculus students know that saddle points are also common. Imagine the "middle" of a saddle (the kind you put on a horse). Looking to the sides where the legs drape down, the middle point seems like a maximum, while looking ahead or behind, the middle

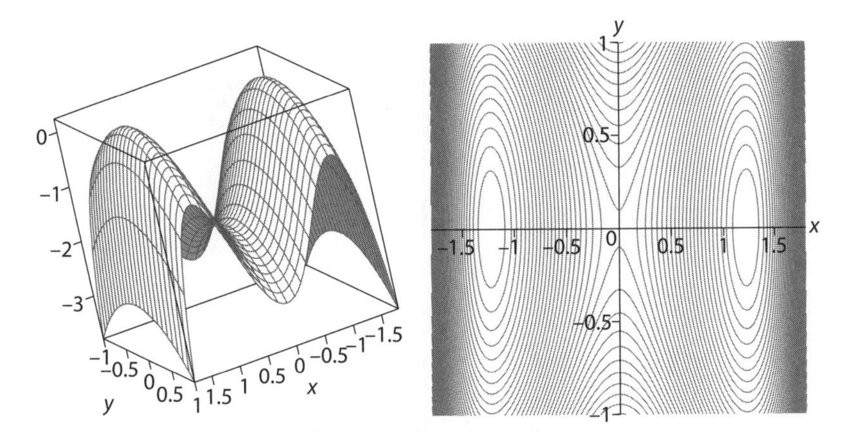

FIGURE 3.23: The surface and contours for $f(x, y) = -(x^2 - 1)(x^2 - 2) - y^2$.

point seems like a minimum. A saddle point can also be visualized with stackable potato chips.

The function $f(x, y) = -(x^2 - 1)(x^2 - 2) - y^2$ has two local maxima and one saddle point (figure 3.23). You could picture this as a landscape with two peaks and a mountain pass running between them. To travel from one side of a mountain range to the other side while staying as low as possible, you pass through the saddle point, the maximum along that path. On the other hand, to travel from one peak to the other while staying as high as possible, the saddle point is encountered again. This time the saddle point is the lowest point along the path.

Let's get back to the question about three critical points. The proper question asks whether there can be a function with two local maxima but no other critical points. Surprisingly, such a function exists (figure 3.24). An example is

$$f(x, y) = -(x^2 - 1)^2 - (x^2 y - x - 1)^2$$

The two local maxima are at the points $(1, 2)$ and $(-1, 0)$. Since the function $f(x, y) \leq 0$ for all (x, y) points—do you see why?—and f equals zero at these two points, this forces them to be local maxima. But there are no other points where the function has a horizontal tangent plane. Far from the origin, the tangent plane may be almost flat but

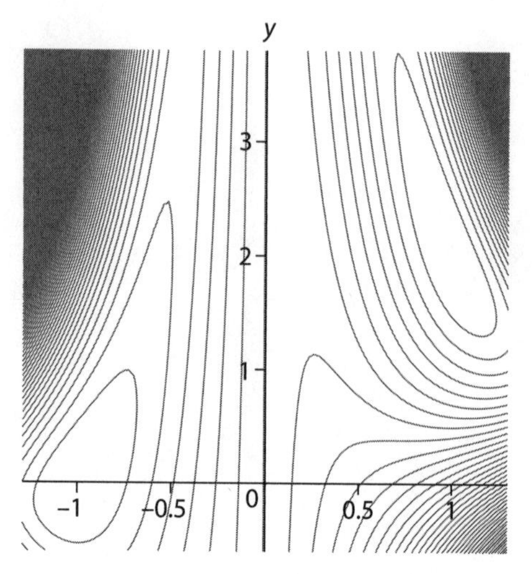

FIGURE 3.24: Contours for $f(x, y) = -(x^2 - 1)^2 - (x^2 y - x - 1)^2$.

not quite. In this sense, we say that the function has a "critical point at infinity." This could be explored further, but do you really want to ride something wilder than the roller coaster?

Sums of Cubes

A classical formula shows that the sum of the first n cubes has a beautiful form:

$$1^3 + 2^3 + \cdots + n^3 = (1 + 2 + \cdots + n)^2 \tag{3.6}$$

Even more interesting is that equation (3.6) can be generalized. Let $\tau(n)$ denote the number of divisors of the number n. For example, $\tau(45) = 6$ since the divisors of 45 are $\{1, 3, 5, 9, 15, 45\}$. The more enticing identity is

$$\sum_{d|n} \tau^3(d) = \left(\sum_{d|n} \tau(d) \right)^2 \tag{3.7}$$

where each sum is over all the numbers d that are the divisors of n. Let's test drive this for $n = 45$ again. Table 3.3 looks at the sum of divisors of d where d is itself a divisor of n:

Table 3.3
Divisors of 45

d	*Divisors of d*	$\tau(d)$
1	{1}	1
3	{1,3}	2
5	{1,5}	2
9	{1,3,9}	3
15	{1,3,5,15}	4
45	{1,3,5,9,15,45}	6

Then

$$\begin{aligned}\sum_{d|45} \tau^3(d) &= \tau^3(1) + \tau^3(3) + \tau^3(5) + \tau^3(9) + \tau^3(15) + \tau^3(45) \\ &= 1^3 + 2^3 + 2^3 + 3^3 + 4^3 + 6^3 \\ &= 324 \\ &= (1+2+2+3+4+6)^2 \\ &= (\tau(1) + \tau(3) + \tau(5) + \tau(9) + \tau(15) + \tau(45))^2 \\ &= \left(\sum_{d|45} \tau(d)\right)^2\end{aligned}$$

By taking $n = 2^m$ in equation (3.7), this identity simplifies to become equation (3.6).

Approximating Decay

The expression "exponential decay," used to mean rapid decline, has a precise mathematical meaning. Often referring to population levels, it means the rate of decay is proportional to the current size of the population. Letting $p(t)$ represent the size of the population at time t, this statement can be quantified by the differential equation $p'(t) = Kp(t)$ for some proportionality constant K. The constant K is negative (if it were positive, we'd be talking about exponential growth). The functions satisfying this equation take the form $p(t) = p_0e^{Kt}$, where p_0 is the population size at $t = 0$. This equation is also used to model radioactive decay.

As all good calculus students know, exponential decay eventually shrinks faster than any rational function. How well can we approximate the exponential function e^{-x} on the interval $[0, \infty)$ with a rational function? Let R_n denote the set of all rational functions that are reciprocals of polynomials and whose degree is at most n. For each such rational function $f(x)$, one could find the largest difference between $f(x)$ and e^{-x}. Now over all the possible rational functions, how small can we keep this largest difference? This can be quantified with the function $\lambda_n = \inf_{f \in R_n} \sup_{x \geq 0} |e^{-x} - f(x)|$. As n gets larger, we have more rational functions at our disposal; hence, the value of λ_n should decrease. A curious result connected to the number 3 is

$$\lim_{n \to \infty} \lambda_n^{1/n} = \frac{1}{3}$$

4

The Number Four

There are only four people who knew what the Beatles were about anyway.

—Paul McCartney

The thing is, we're all really the same person. We're just four parts of the one.

—Paul McCartney

The number 4 strikes one as balanced: the number of players in a bridge game, double dates, and the legs of a table. We shall see that in many other scenarios, the number offers the perfect balance, whether it's four colors, four travelers, or four corners. So whether you're lying in a four-poster bed or sitting in the relaxing university quad, enjoy the number 4.

The Four Color Theorem

Maps have always had an aesthetic grip on people. They capture large amounts of information that can be visually assimilated with relative ease. To more readily distinguish between neighboring regions, it makes sense to shade them with different colors. Of course, giving each region its own unique color works, but with a large numbers of regions, making distinctions between these colors may become challenging.

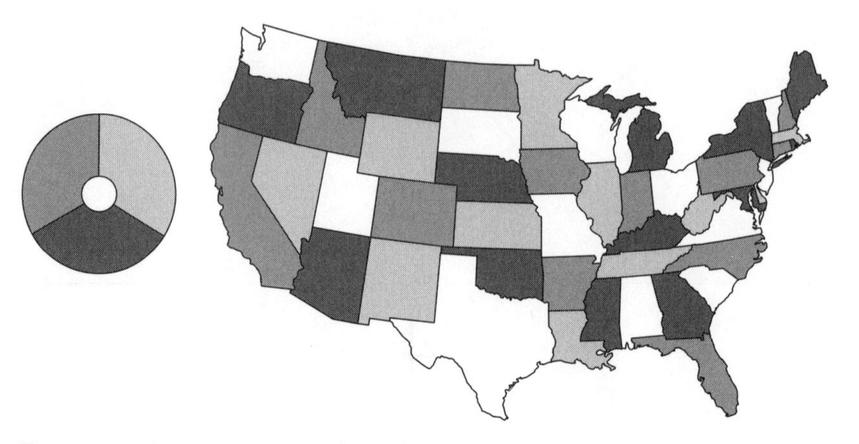

FIGURE 4.1: Four colors are sometimes needed, and four colors suffice.

This leads to the question, "What is the least number of colors needed to color a map so that no two adjacent regions share the same color?" To be clear, if two regions touch only at a point—such as Arizona and Colorado—they can share the same color. A little experimentation shows that at least four colors are needed for some graphs; figure 4.1 shows a scenario where fewer than four colors do not suffice.

The four color problem claims that four colors are sufficient for any map. This assertion was first posed by Francis Guthrie in 1852. The prominent mathematician Augustus De Morgan accelerated the problem's dissemination. The history of the problem and its eventual solution has had its fair share of colorful characters, as well as controversy. Alfred Kempe contributed some of the initial ideas that would eventually be used in the proof. Kempe actually claimed that he had proved the whole theorem in 1879, but he later suffered a fate most researchers would consider tantamount to meeting the Four Horsemen of the Apocalypse: in 1890, Kempe's published proof was found to be flawed. While errors can creep into scientific results for a host of reasons (experimental error, incorrect statistical analyses, corrupted data, etc.), mathematicians typically have only their own faulty thought processes to blame. A mathematical claim is based on logic and is not subject to taste or technological change.

Percy Heawood—the person who found the flaw in Kempe's proof—did more than just uncover an error. He showed that Kempe's argument could be modified to imply that any map is *five*-colorable. One could not be blamed for hoping that since a manageable proof for five colors was found, proving that each map was four-colorable shouldn't be too difficult. Several decades would pass before a proof could be offered, and even then it was not what anyone would have expected.

In 1976, Kenneth Appel and Wolfgang Haken announced that they had a proof of the Four Color Theorem. We write "announced" because many mathematicians at that time—and even some now—would not accept the proof. Why not? Appel and Haken showed that proving the theorem could be boiled down to checking it for 1,936 (later reduced to 1,482) specific cases. They then used a computer to check each case. If all the details were made explicit, their write-up would explode to hundreds of pages and 10,000 diagrams. Why do some mathematicians feel uneasy about this proof? In a nutshell, they don't trust the computer. Mathematics distinguishes itself from most other scholarly disciplines because each step in an argument should be logically watertight. The logic doesn't depend on experimental observation, expert opinion, or popular consent. A logical argument can stand forever. Mathematicians take great comfort in the surety of their proofs. The idea of farming out part of a proof to a machine can seem distasteful, particularly to a purist. Although Appel and Haken explained the structure in their proof—that is, why a finite number of cases was sufficient—and explained what the computer was checking (which could be replicated by anyone), many mathematicians still felt somewhat bamboozled. Besides the concerns about trusting the computer, there was the desire, if not the expectation, that this simply stated conjecture should have an equally simple solution. Of course, everyone would prefer a shorter traditional proof, but such a neatly packaged result may not exist. We may have to content ourselves with a long proof to a simple, elegant problem.

The chasm of difficulty between showing that four colors suffice and five colors suffice is also demonstrated in map-coloring algorithms. Knowing that four colors are enough to color a specific map is one thing, but finding the coloring is another matter. In 1996, an algorithm was developed to determine how to five-color a region. The computational requirements for this algorithm are proportional to the number of regions. In contrast, an algorithm for four-coloring a map is proportional to the square of the number of regions and is thus much more costly.

FIGURE 4.2: Tennis ball and baseball.

The Tennis Ball Theorem

Have you ever noticed an interesting pattern on a standard tennis ball? The seam divides the ball into two identical dumbbell-shaped pieces. A baseball has the same pattern (figure 4.2). If you trace your finger along the seam of the ball, you may note that there are four points where your motion switches from "bending left" to "bending right," or vice versa. This is no coincidence. Those special places where the curve is neither bending right nor bending left are called *points of inflection*. The Tennis Ball Theorem claims that if a smooth, closed curve divides a sphere into two equal areas, then there are at least four points of inflection on the curve. The trivial situation of taking the equator around the sphere yields an inflection point at *every* point of the curve. Note that the theorem may not work if the equal area assumption is dropped. For example, if the curve is, say, the Tropic of Capricorn, the curve always bends in the same direction, so there are no inflection points.

Sum of Squares Identities

If n is a positive integer, what is the fewest number of squares needed that sum to n? When n is a prime, sometimes only two squares are needed. This is captured in the beautiful result known as Fermat's Two-Squares Theorem: If p is a prime number where $p \equiv 1 \bmod 4$, then

$p = x^2 + y^2$ for some integers x and y. Examples of Fermat's result include $13 = 2^2 + 3^2$, $41 = 4^2 + 5^2$, and $137 = 4^2 + 11^2$.

Fermat's Two-Squares Theorem can be easily extended: if the prime factors of n are all congruent to 1 mod 4, then n is the sum of two squares. Why? Suppose that $n = p_1 p_2 \cdots p_k$, a product of primes each congruent to 1 mod 4. Each of these primes can be written as a sum of two squares. By using the Fibonacci–Brahmagupta formula, namely

$$(a_1^2 + a_2^2)(b_1^2 + b_2^2) = (a_1 b_1 + a_2 b_2)^2 + (a_1 b_2 - a_2 b_1)^2,$$

the product $p_1 p_2$ is also the sum of two squares. Multiplying this result by p_3 also gives a sum of two squares. This process can be repeated indefinitely until n is a sum of two squares.

Of course, many numbers (prime or otherwise) cannot be written as the sum of two squares. The number seven requires four squares: $7 = 2^2 + 1^2 + 1^2 + 1^2$. Can every positive integer be written as the sum of four squares? Yes! This is the claim of Lagrange's Four-Square Theorem. An important tool—reminiscent of the Fibonacci–Brahmagupta formula—is Euler's four-square identity:

$$\begin{aligned}&(a_1^2 + a_2^2 + a_3^2 + a_4^2)(b_1^2 + b_2^2 + b_3^2 + b_4^2)\\ &= (a_1 b_1 - a_2 b_2 - a_3 b_3 - a_4 b_4)^2 + (a_1 b_2 + a_2 b_1 + a_3 b_4 - a_4 b_3)^2\\ &\quad + (a_1 b_3 - a_2 b_4 + a_3 b_1 + a_4 b_2)^2 + (a_1 b_4 + a_2 b_3 - a_3 b_2 - a_4 b_1)^2\end{aligned}$$

A result like Fermat's Two-Squares Theorem shows that any prime can be written as the sum of four squares. By using Euler's identity, Lagrange's Theorem follows.

Rearranging Four Pieces

A lovely puzzle attributed to Henry Dudeney in 1907 asks how to decompose an equilateral triangle into four pieces that can be rearranged to form a square (figure 4.3). Dudeney's puzzle is a special instance of a much more general and amazing mathematical result.

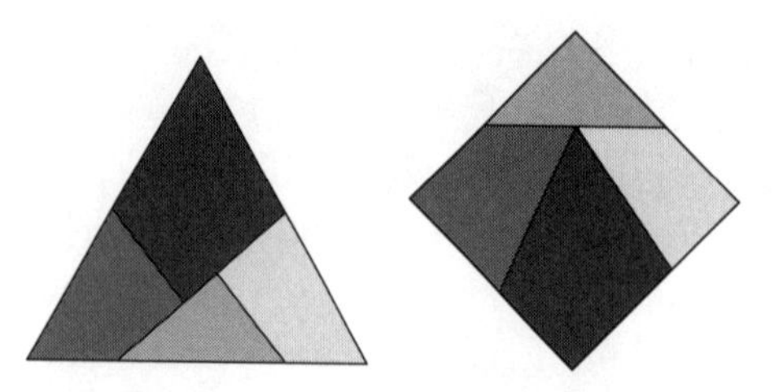

FIGURE 4.3: Changing an equilateral triangle into a square.

The Wallace–Bolyai–Gerwien Theorem claims that any two polygons of equal area are *equidecomposable*. This means that one of the polygons can be minced into a finite number of pieces and rearranged to form the other polygon.

The proof of this remarkable theorem can be broken down into a few steps. First, note that if we can dissect any polygon so that it can be rearranged into a square of the same area, we will be finished. To accomplish this dissection, first cut the polygon into triangles. Next, cut each triangle into two right triangles. Now we start the reconstruction. Take each right triangle and make one cut so that the two pieces form a rectangle. The hardest step is to show that a rectangle can be dissected in such a way that it can be reconstructed into any other rectangle with the same area. This can be used to stack appropriately sized rectangles together to form the square.

While this proof offers an algorithm for transforming one polygon into another with the same area, the number of pieces produced in the process may pile up quickly. This makes Dudeney's puzzle even more striking; only four pieces are needed.

Caution must be taken, however, when making claims about equidecompositions. Figure 4.4 shows a square that is dissected into four parts and seemingly rearranged into a rectangle. By comparing areas, it seems that $64 = 65$. Can you find the problem? The solution is in Chapter 10.

Ducci Sequences

Let's play a simple game. Start with four numbers, say 2, 5, 7, and 13. Take the absolute values of the differences between each pair of

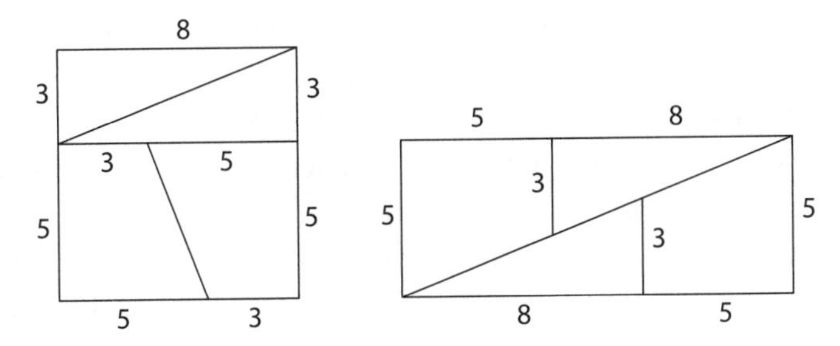

FIGURE 4.4: What is the mistake in these figures?

neighbors—think "wrap around" so that 2 and 13 are neighbors—to produce four new numbers: $(3, 2, 6, 11)$. Iterating this process produces

$$\begin{aligned}(3,2,6,11) &\to (1,4,5,8) \to (3,1,3,7) \to (2,2,4,4) \\ &\to (0,2,0,2) \to (2,2,2,2) \to (0,0,0,0) \qquad (4.1)\end{aligned}$$

Of course $(0, 0, 0, 0)$ iterates to itself. There is a beautiful convergence theorem for this process: if one starts with *any* four integers, the corresponding sequence will arrive at $(0, 0, 0, 0)$ in a finite number of iterations. Since this phenomenon is attributed to E. Ducci, most papers refer to this process (and variations thereof) as *Ducci sequences*.

The earliest documented observation of these sequences is in an obscure Italian paper from 1937. Since this publication was not widely read, the convergence theorem about Ducci sequences has been rediscovered and reproven several times. Why all this rediscovery? First, the process is so simple that a child could have made the observations. Second, even if you were convinced that someone had already invented this process, how could you find it? While there are some tools available for searching the mathematical literature, looking for an idea that is bare of any name or sophisticated properties is difficult. Until Ducci sequences become a well-known part of mathematical knowledge, they are destined to be regularly rediscovered.

The reader may find the Ducci process to be similar to the $3x + 1$ problem mentioned in the last chapter. The British mathematician Sir Bryan Thwaites—whom you may remember claimed to have invented the $3x + 1$ problem—wrote a short note about these two problems called "Two Conjectures, or How to Win £1100." Unaware that the Ducci problem had been solved long ago, he offered £100 for its resolution, while the sterner looking $3x + 1$ problem had a £$1,000$ prize attached to it. Like a goldfish in a piranha tank, the first prize was gobbled up instantly. The second prize, of course, has languished to this day. One observation that demonstrates the difference between these two problems is the change of size in their iterations. For a Ducci process, the maximum number in any 4-tuple never gets larger (and in general goes down). In the example (4.1), the maximums are

$$13 \to 11 \to 8 \to 7 \to 4 \to 2 \to 2 \to 0$$

In contrast, iterations in the $3x + 1$ problem may climb toward the Moon or come crashing down to the depths. This apparent lack of patterns is what makes the $3x + 1$ problem so hard.

Applying the Ducci process to sets of three numbers does not give the same result. For example, note that

$$(0, 1, 1) \to (1, 0, 1) \to (1, 1, 0) \to (0, 1, 1) \tag{4.2}$$

so the process loops and never reaches $(0, 0, 0)$. Does the number of terms give us information about the long-term dynamics? A general theorem states that if the number of terms is a power of 2, then any initial sequence will reach all zeros in a finite number of iterations. If the number of terms is not a power of two, there is always a starting set of numbers that loops like in example (4.2).

Even though Ducci sequences starting with four numbers reach all zeros, there is no upper bound on how many iterations it takes to get there. To see this, consider the Tribonacci numbers (sorry, Fibonacci) defined by

$$t_n = t_{n-1} + t_{n-2} + t_{n-3}$$

$$t_0 = 1, t_1 = 1, t_2 = 2$$

The sequence starts as $1, 1, 2, 4, 7, 13, 24, 44, 81, 149, 274$. Applying the Ducci process to $(t_n, t_{n-1}, t_{n-2}, t_{n-3})$, a beautiful structure is revealed: three iterations produce a shifted version $2 \cdot (t_{n-2}, t_{n-3}, t_{n-4}, t_{n-5})$. Starting with $(t_{104}, t_{103}, t_{102}, t_{101})$, for example, 150 iterations produce a shifted version of $2^{50}(t_4, t_3, t_2, t_1)$, and six more iterations take us to $(0, 0, 0, 0)$. By choosing a larger value for n, we can make the number of iterations needed to reach $(0, 0, 0, 0)$ as large as possible.

Strange things can happen, however, if our numbers are not restricted to integers. Just as the quotients of Fibonacci numbers can be made arbitrarily close to the Golden Ratio, quotients of Tribonacci numbers can be made close to the irrational number $q \approx 1.839$, a solution to $q^3 - q^2 - q - 1 = 0$. This implies that $(t_n, t_{n-1}, t_{n-2}, t_{n-3})$ can be made close to a scaled version of $(q^3, q^2, q, 1)$ by choosing large values of n. This irrational 4-tuple iterates in an interesting way:

$$\begin{aligned}(q^3, q^2, q, 1) &\to (q-1) \cdot (q^2, q, 1, q^3)\\ &\to (q-1)^2 \cdot (q, 1, q^3, q^2)\\ &\to (q-1)^3 \cdot (1, q^3, q^2, q)\\ &\to (q-1)^4 \cdot (q^3, q^2, q, 1)\end{aligned}$$

In other words, four iterations of $(q^3, q^2, q, 1)$ produce a scaled version of itself! Since the scaling factor $(q - 1)^4$ is less than 1, these iterations also approach all zeros, but now it takes an infinite number of iterations.

Euler's Sum of Powers Conjecture

Euler is credited with proving the cubic case of Fermat's Last Theorem (FLT), that is, there are no positive integer solutions to $x^3 + y^3 = z^3$. In the spirit of FLT, Euler conjectured that the equation

$$a_1^k + a_2^k + \cdots + a_n^k = b^k$$

has no positive integer solutions if $n < k$. Euler's Conjecture would encapsulate the two earlier results since the case $k = 3$ is the cubic case of FLT and the case $n = 2$ is all of FLT.

Euler's Conjecture remained open for centuries, and, like Fermat's Last Theorem, was considered a mountain too steep to climb. It was a classical problem that was easy to state but seemed to lack the tools to solve it. So it was with surprise that in 1966 a counterexample was found by L. J. Lander and T. R. Parkin for the $k = 5$ case. Using a brute force computer search, they found

$$27^5 + 84^5 + 110^5 + 133^5 = 144^5$$

Another counterexample,

$$85,282^5 + 28,969^5 + 3,183^5 + 55^5 = 85,359^5$$

was found by Jim Frye in 2004.

After encountering these solutions for $k = 5$, it's natural to ask if counterexamples to the conjecture hold for $k = 4$. In 1986, Noam Elkies used tools from a modern area of research in number theory, namely elliptic curves, to find a counterexample:

$$2,682,440^4 + 15,365,639^4 + 18,796,760^4 = 20,615,673^4$$

In fact, Elkies' approach produces infinitely many solutions. Discovered when Elkies was 20, this result (and others) cemented his stature as one of the brightest mathematicians of his generation. He has also attracted attention for his piano performances, musical compositions, and mastery at solving chess problems. At age 26, he became the youngest full professor at Harvard University (figure 4.5).

Curiously, Don Zagier, an already established mathematical star, was independently working on the same problem at the same time. Like Elkies, Zagier also completed his Ph.D. when he was very young. He has made seminal contributions in number theory and has positions at two institutions, the Max Planck Institute for Mathematics in Bonn, Germany, and the Collège de France in Paris. Zagier is also a polyglot, fluent in many languages, including English, German, French, Dutch, Italian, and Russian.

But let's get back to Euler's Conjecture for $k = 4$. Zagier's effort started after speaking with a colleague at Berkeley in 1986, sparking a new idea. After having made good progress with this new approach, Zagier left for a two-month trip to Moscow. He reached a point in this research where he needed a computer, but his computational options in the politically thawing USSR were limited. Armed with only a pocket

FIGURE 4.5: Noam Elkies (top) and Don Zagier (bottom).

calculator—no one traveled with anything more powerful in those days—he could not complete the necessary calculations. No worries, though. He would be back in Bonn soon enough. This conjecture had been open for more than 200 years; nobody was going to scoop him.

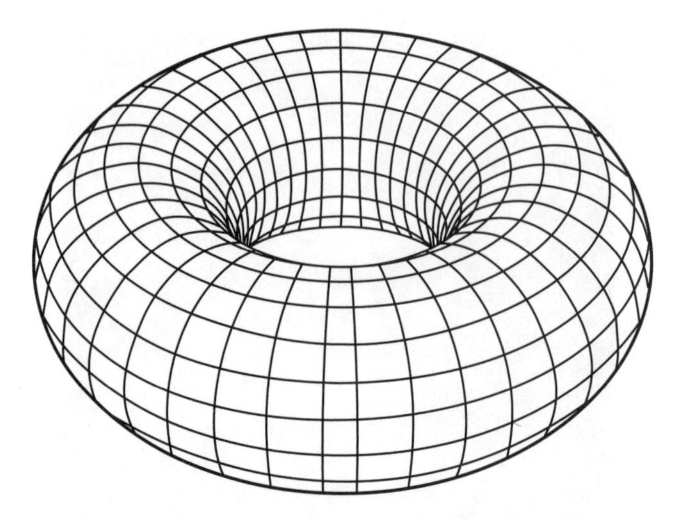

FIGURE 4.6: A torus and two coverings with circles.

Wrong. Upon Zagier's return, a colleague excitedly told him that Euler's Conjecture had been settled by the young American Noam Elkies just days earlier. Zagier rushed to his computer and, within seconds, found a solution. So a problem of Euler's that had stood open for more than two centuries was solved completely independently by two people (figure 4.5) just days apart.

Villarceau Circles

In Chapter 3, we saw that the torus has circles on its surface. Families of circles can be used to cover the torus completely in two different ways (figure 4.6). Another way to view this phenomenon is that through each point on the torus there are two circles that pass through it and lie completely on the surface. Being greedy, we could ask, "Are there any more circles lying on the torus going through a specified point?" Surprisingly, the answer is yes. There are two more circles that go through each point called *Villarceau circles*. To construct these new circles, start by making a vertical plane that passes through the given point and the center of the torus. Now tilt the plane—while still going through these two points—until it is tangent to the torus at yet another point. This plane intersects the torus in a circle (figure 4.7). There are two tilting angles that produce this effect, so two new circles have been

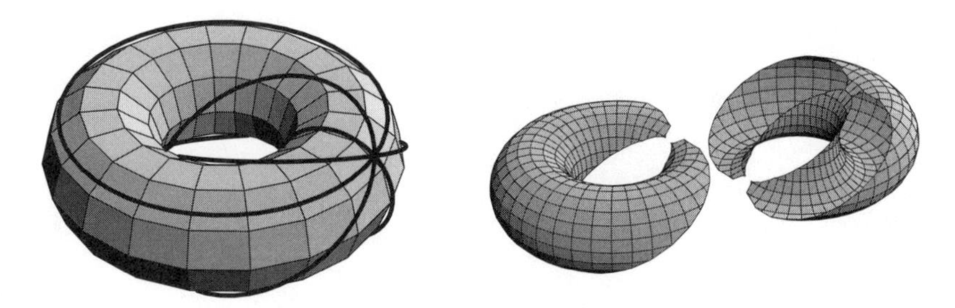

FIGURE 4.7: Villarceau circles.

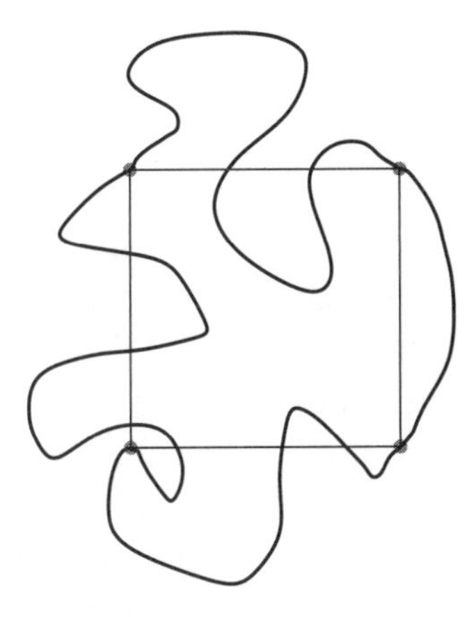

FIGURE 4.8: An inscribed square.

formed. In total, each point on the torus has four circles passing through it that lie completely on the surface.

The Inscribed Square Problem

The claim sounds easy enough: on any simple closed curve there exist four points that are the vertices of a square. Figure 4.8 demonstrates an example.

Some curves (such as circles) admit infinitely many inscribed squares, whereas a noncircular ellipse or a triangle with an interior angle greater than 90° admit only one. The Inscribed Square Problem, proposed by Otto Toeplitz in 1911, is sometimes called the *Toeplitz Problem*. It has been shown that if the curve is smooth enough, then there is an inscribed square. Unfortunately, a general proof has been elusive.

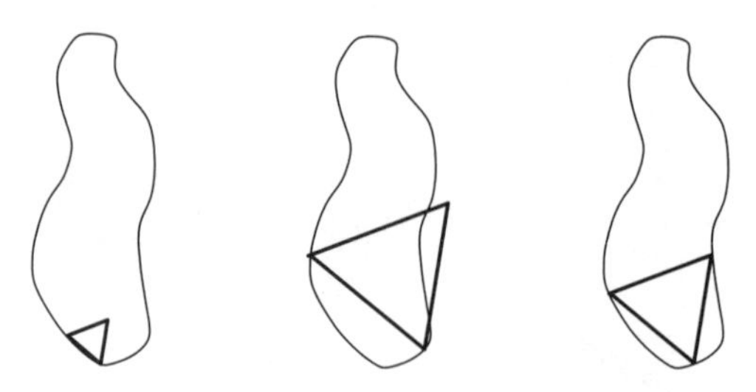

FIGURE 4.9: Constructing an inscribed equilateral triangle.

So that we walk away with more than nothing, it is easy to show that every simple closed curve admits an inscribed equilateral triangle. Start with a small equilateral triangle in the interior of the curve. By shifting and rotating the triangle, it can be oriented so that two of its vertices lie on the curve while the third vertex remains in the interior. Now start moving the two points away from each other, while staying on the curve, and changing the third point so that an equilateral triangle is maintained at every instant. If the two points are moved so that they are as far apart as possible, the third point can no longer be in the interior. Since the third point is outside the curve, it must have passed though the curve at some point in time. In that instant, we would have an equilateral triangle with each point on the curve. Figure 4.9 shows this Goldilocks scenario; one triangle is too small, one is too large, and one is just right.

Regular Polygons on a Computer Screen

How can one produce perfect polygons on a computer screen? By "perfect" we mean each vertex of the polygon is centered exactly in the center of a pixel. Unfortunately for computer graphics, this is impossible for any regular polygon except for a square.

Figure 4.10 gives the essence of why this fails for a regular pentagon. Assume that the vertices of the pentagon have integer coordinates. Rotating each side by 90°—see the dashed lines in the figure—creates an enclosed pentagon whose vertices also have integer coordinates. Of course, one can iterate this procedure, reaching a contradiction when the pentagon is sufficiently small. This same process generalizes to any regular polygon with more than four sides. Lastly, we also get

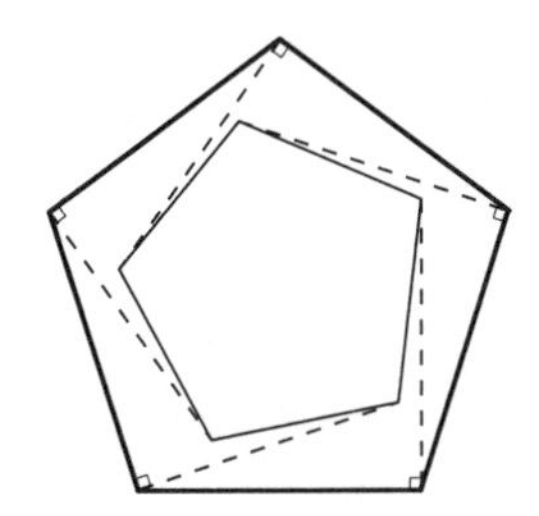

FIGURE 4.10: Regular pentagons cannot have integer coordinates.

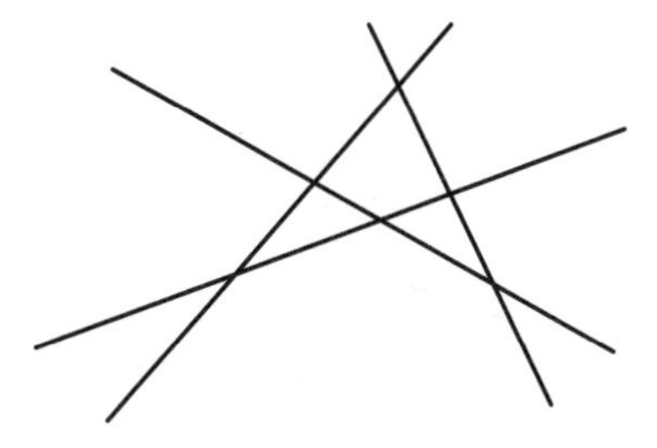

FIGURE 4.11: Four lines in general position.

the same result for equilateral triangles at no cost. Why? If there was an equilateral triangle with integer coordinates, there must also be a regular hexagon with integer coordinates, but we have shown that this is impossible.

The Four Travelers Problem

Imagine four straight roads where no two are parallel and no three intersect in one point (this is sometimes called *general position*) (figure 4.11). On each road there is a traveler who moves at a constant speed (in general, the four speeds will be different). By some wild coincidence, the position and speed of the travelers works out just right so that travelers 1 and 2 meet each other and the other travelers at their respective crossings. The claim is that travelers 3 and 4 will also meet at their crossing.

While this problem can be solved with a modest dose of algebra, the reader will be spared the details. However, we will elaborate on some interesting technical aspects of the solution. Let t_1 denote the time when traveler 1 and traveler 2 meet, t_2 when 1 and 3 meet, t_3 when 1 and 4 meet, t_4 when 2 and 3 meet, t_5 when 2 and 4 meet, and

Table 4.1
Positions of the Four Travelers

Traveler	(x, y)-*position*
1	$(0, 3t)$
2	$(6t, 0)$
3	$(12t - 12, 6 - 3t)$
4	$(-2t - 4, 4t + 2)$

Table 4.2
Intersection Times of the Four Travelers

k	1	2	3	4	5	6
t_k	0	1	-2	2	−1/2	4/7

t_6 when 3 and 4 meet. There is a relationship between these six points in time via a mysterious function. If we let

$$h(t_1, t_2, t_3, t_4, t_5, t_6) = -t_6t_2t_5 - t_6t_1t_3 + t_6t_1t_5 + t_6t_4t_3 + t_6t_2t_1 - t_6t_4t_1 + t_5t_3t_2 + t_5t_4t_2 - t_4t_3t_2 + t_4t_3t_1 - t_5t_4t_3 - t_2t_1t_5$$

then

$$h(t_1, t_2, t_3, t_4, t_5, t_6) = 0 \tag{4.3}$$

If any five of these times are specified, the sixth one can be easily found. As an example, suppose we specify the location of each traveler at a general time t (table 4.1). Some algebra can be used to calculate the intersection times; see table 4.2. In a similar way, one could analyze the most general problem and produce equation 4.3.

Although the expression for h seems complicated, it possesses a rich, latent structure. In retrospect, this is not surprising given the structure in the Four Travelers Problem; the invariant properties of the problem force some complexity in the solution. For example, one would expect that if time was measured at a different rate—say in hours instead of minutes—this should not alter the result of the problem. This claim is borne out by the identity

$$h(ct_1, ct_2, ct_3, ct_4, ct_5, ct_6) = c^3 h(t_1, t_2, t_3, t_4, t_5, t_6)$$

for any constant c. Another algebraic property of the function h is encountered if time is shifted. For example, we could modify the quantities so that $t = 0$ corresponds to noon instead of eight in the morning, shifting t by a value of four. This change would shift all the intersection times by the same amount and equation (4.3) would still be satisfied. This generates the algebraic surprise

$$h(t_1 + s, t_2 + s, t_3 + s, t_4 + s, t_5 + s, t_6 + s) = h(t_1, t_2, t_3, t_4, t_5, t_6)$$

for any value of s. This identity can be verified by expanding and simplifying the left side. Lastly, if we switch the labels on lines 1 and 2, this means that t_1 and t_6 remain unchanged but t_2 and t_4 swap, and t_3 and t_5 swap. This permutation of crossing times is captured in the formula

$$h(t_1, t_4, t_5, t_2, t_3, t_6) = -h(t_1, t_2, t_3, t_4, t_5, t_6) \tag{4.4}$$

There are five other ways to swap two roads that produce five other equations:

$$\begin{aligned} -h(t_1, t_2, t_3, t_4, t_5, t_6) &= h(t_4, t_2, t_6, t_1, t_5, t_3) \\ &= h(t_5, t_6, t_3, t_4, t_1, t_2) \\ &= h(t_2, t_1, t_3, t_4, t_6, t_5) \\ &= h(t_3, t_2, t_1, t_6, t_5, t_4) \\ &= h(t_1, t_3, t_2, t_5, t_4, t_6) \end{aligned}$$

The Four Travelers Problem shows that buried under an interesting geometric result lie some beautiful algebraic equations.

The Four Exponentials Conjecture

As we saw in Chapter 1, the real numbers can be divided into two classes: the rationals and the irrationals. Since the rationals are countable and the irrationals are uncountable, there is a sense in which "most" numbers are irrational. If you throw a dart at the real line, there is a 100% chance—from a probabilistic point of view—that it would strike an irrational number.

The irrational numbers can be further subdivided. A number x is said to be *algebraic* if there is a polynomial p of one variable with integer coefficients such that $p(x) = 0$. For example, $x = 3^{1/4}$ is algebraic since $p(x) = 0$ when $p(x) = x^4 - 3$. Numbers that are not algebraic—these are called *transcendental* numbers—are a subclass of the irrationals. Just as the rational numbers are countable, a similar argument can be made showing that the algebraic numbers are also countable. This means that the transcendental numbers also hog the real line.

Proving that a suspected number is irrational is often a difficult feat, but showing that it is transcendental requires more muscle than a four-by-four truck. In 1761, Johann Heinrich Lambert proved that π is irrational, but the first proof that π is transcendental was produced only in 1882. Since almost all numbers are transcendental, it would seem reasonable to have a simple test that shows a given number is transcendental. Perhaps the simplest stated result in this direction is the Gelfond–Schneider Theorem. This result claims that if a and b are algebraic (except for 0 and 1), and b is irrational, then a^b is transcendental. For example, this theorem easily shows that $2^{\sqrt{2}}$—dubbed the Gelfond–Schneider constant—is transcendental.

Sometimes this theorem is used in the negative sense to produce an interesting result. One of the most beautiful formulas in mathematics is Euler's formula $e^{\pi i} = -1$. Rewrite this as $(e^{\pi})^i = -1$ and consider the Gelfond–Schneider Theorem with $a = e^{\pi}$ and $b = i$. Since the right side of the formula, -1, is *not* transcendental, the conditions of the theorem must break down at some place. The exponent $b = i$ is irrational and algebraic since $i^2 + 1 = 0$. Furthermore, $a = e^{\pi}$ is neither zero nor one. The only option left is that e^{π} is not algebraic, that is, it is transcendental.

We finally arrive at a claim connecting transcendental numbers and 4, the unresolved Four Exponentials Conjecture. Suppose that x_1, x_2 and y_1, y_2 are pairs of complex numbers that are linearly independent over the rationals. This means that if there are rational numbers p, q such that $px_1 + qx_2 = 0$ or $py_1 + qy_2 = 0$, then $p = q = 0$. The conjecture claims that at least one of the four numbers

$$e^{x_1 y_1},\ e^{x_1 y_2},\ e^{x_2 y_1},\ e^{x_2 y_2}$$

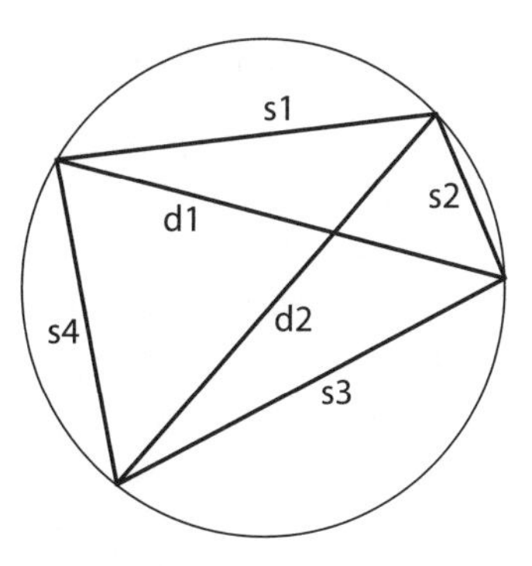

FIGURE 4.12: Ptolemy's Theorem.

is transcendental. This problem was first mentioned in print by Theodor Schneider in 1957.

Concentric Quadrilaterals

One of the gems of ancient Greek geometry is Ptolemy's Theorem. Suppose that a quadrilateral is concentric, that is, its four vertices lie on a circle. If the side lengths, in order, are s_1, s_2, s_3, and s_4 and the diagonal lengths are d_1 and d_2—see figure 4.12—then Ptolemy's Theorem states

$$s_1 s_3 + s_2 s_4 = d_1 d_2 \tag{4.5}$$

Moreover, the converse of Ptolemy's Theorem also holds true; if a quadrilateral satisfies equation (4.5), then it must be concentric. This gives an easy way to check if a quadrilateral is concentric. Well, sometimes it's easy. From a computational perspective, calculating the six lengths involves square roots, so if numerical approximations have to be made, it's likely that the calculation will be slightly off. Moreover, equation (4.5) requires pairing up opposite side lengths of the quadrilateral. With a picture, this may be straightforward, but if one wants to program a computer to put the points in cyclic order, it is not obvious how this would be done.

To address this challenge, let's represent the quadrilateral in a different form. Consider the four vertices of a quadrilateral as complex numbers a, b, c, d in the complex plane. Equation (4.5) can be rewritten as

$$|a-b||c-d|+|b-c||d-a|=|a-c||b-d|$$

Another form to characterize when four points are concentric involves a quantity commonly used in complex analysis, the *cross ratio*. The cross ratio, denoted by $cr(a, b, c, d)$, is defined as

$$cr(a, b, c, d)=\frac{(a-c)(b-d)}{(b-c)(a-d)}$$

Although the cross ratio has been around for a long time, it was only studied extensively in the nineteenth century. Among this strange-looking quantity's virtues is a new way to characterize when four points are concentric: the quadrilateral formed by the four complex numbers a, b, c, d is concentric if and only if the quantity $cr(a, b, c, d)$ is a real number. Using this characterization to prove that a quadrilateral is concentric avoids the two pesky concerns mentioned earlier: no square roots are involved, and the points a, b, c, d do not need to be ordered.

For those who like algebra, an interesting formula that connects these two approaches is

$$\begin{aligned}
&[|a-c||b-d|-|a-b||c-d|-|b-c||d-a|]\\
&\times[|a-c||b-d|-|a-b||c-d|+|b-c||d-a|]\\
&\times[|a-c||b-d|+|a-b||c-d|-|b-c||d-a|]\\
&\times[|a-c||b-d|+|a-b||c-d|+|b-c||d-a|]\\
&=\left[(a-b)\overline{(b-c)}(c-d)\overline{(d-a)}-\overline{(a-b)}(b-c)\overline{(c-d)}(d-a)\right]^2\\
&=-4|a-d|^4|b-c|^4\,(Im(cr(a, b, c, d)))^2
\end{aligned}$$

The expression $Im(z)$ refers to the imaginary part of z. This equation holds for *any* complex numbers a, b, c, d. The left side is the product of four large expressions. If any of the first three expressions equal zero, such an equation is equivalent to equation (4.5) (depending on

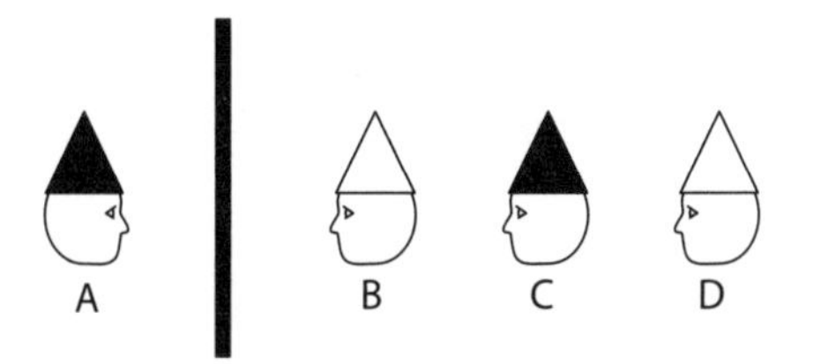

FIGURE 4.13: The Four Hats Problem.

the ordering of the points). The fourth expression, composed of only positive quantities, is never zero. The right side of equation (4.5) equals zero when the cross ratio is real (we're assuming that the four points are distinct). This shows the equivalency of the Ptolemy and cross ratio approaches.

The Four Hats Problem

To punish four naughty boys, their teacher decides to keep them in from recess where they must sit quietly in the classroom. To offer them a way out, the teacher devises a problem to solve. The boys are seated in a line with three facing the same way but the fourth facing back toward the others. A panel is placed between the first three and the fourth; see figure 4.13. The teacher then places four hats on the boys' heads, two black and two white. Each boy knows that there are two hats of each color, but he only knows who is wearing what color by what he can see: C knows B's color and D knows B and C's colors. No one can see the color of his own hat.

The teacher announces, "I'm giving you a way to have recess, but you first need to solve a puzzle. Within the next five minutes, one of you must announce what color hat he is wearing. If you are correct, you all can go to recess. If anyone makes a false claim, you must all stay in. There is no communication allowed between you." After one minute, one of them announces the color of his hat. Who was it, and how is he sure that he is correct? The solution is in Chapter 10.

A variant of the problem has three white hats and one black hat and students B, C, and D are situated so that they can all see each other. Student A still cannot see the others. How is it that after one minute one student knows which color hat he is wearing?

5

The Number Five

I think there is a world market for maybe five computers.
—alleged 1943 statement by Thomas J. Watson

Five is the first number that defies being categorized. It embraces order as the number of Platonic solids and is baked into the Rogers–Ramanujan identities, yet it shakes a fist at closed-form solutions to quintic equations or fitting nicely into tessellations. Austistic savant Daniel Tammet asserts that "five is a clap of thunder or the sound of waves crashing against rocks." You will see how it juxtaposes beauty with power.

The Miquel Five Circles Theorem

Starting with a "main circle," build five other circles whose centers are on the main circle where each new circle intersects its two neighbors; see figure 5.1. Now construct five lines through the neighboring inner intersection points. The theorem says that these five lines form a pentagram whose vertices lie on the five satellite circles.

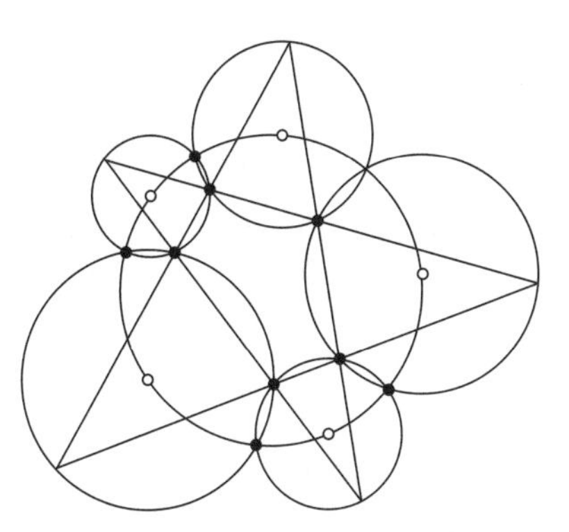

FIGURE 5.1: The Miquel Five Circles Theorem.

The Platonic Solids

The five Platonic solids have mesmerized us since the time of Euclid. The Greeks paired each solid with one of the elements, and Kepler related them to the five extraterrestial planets known at that time. Many mathematicians today encounter the icosahedron's ubiquitous image as the logo for the Mathematical Association of America. So why are there exactly five Platonic solids?

Recall that a Platonic solid is a convex polyhedron whose faces are congruent, regular polygons and that has the same number of edges meeting at each vertex. A cube is a simple example: the faces are congruent squares, and each vertex is met by three edges. To identify the Platonic solids, we need to relate these characteristics. Let e, f and v denote the number of edges, faces, and vertices of the solid. Moreover, let p represent the number of edges surrounding each face and q the number of edges meeting at each vertex. To count the total number of edges, multiply the number of edges around each face by the number of faces. Since each edge is counted for two different faces, we arrive at

$$pf = 2e \tag{5.1}$$

Another way to count the edges is to multiply the number of edges coming into each vertex by the number of vertices, but again, this double-counts since each edge arrives at two different vertices, leading to

$$qv = 2e \tag{5.2}$$

Table 5.1
Characteristics for the Five Platonic Solids

Name	*p*	*q*	*v*	*e*	*f*
Tetrahedron	3	3	4	6	4
Cube	4	3	8	12	6
Octahedron	3	4	6	12	8
Icosahedron	3	5	12	30	20
Dodecahedron	5	3	20	30	12

One more relationship is obtained by "flattening" the solid. Pick a point on a face and imagine stretching the face so that the rest of the solid lies flat on a plane. The face that was opened is now the region exterior to a graph. We thus have Euler's formula from chapter 2:

$$v - e + f = 2 \tag{5.3}$$

Equations (5.1)–(5.3) are used to solve for e, f, and v in terms of p and q:

$$v = \frac{4p}{4 - (p-2)(q-2)}$$

$$f = \frac{4q}{4 - (p-2)(q-2)}$$

$$e = \frac{2pq}{4 - (p-2)(q-2)}$$

Since all the variables are positive, this forces $(p-2)(q-2) < 4$. Such a severe restriction leaves only five possibilities; see table 5.1. These attractive equations show that it's possible to have sensual relationships that are also platonic.

Solving Polynomial Equations

Every high school student learns the dreaded quadratic formula: the solutions to the equation $ax^2 + bx + c = 0$ are

$$x = \frac{-b \pm \sqrt{b^2 - 4ac}}{2a} \tag{5.4}$$

Although this formula is usually shrouded in mystery, its proof is simply a matter of completing the square:

$$ax^2 + bx + c = a\left(x^2 + \frac{b}{a}x + \frac{c}{a}\right)$$

$$= a\left(\left(x + \frac{b}{2a}\right)^2 - \frac{b^2}{4a^2} + \frac{c}{a}\right)$$

$$= a\left(\left(x + \frac{b}{2a}\right)^2 - \frac{b^2 - 4ac}{4a^2}\right)$$

Setting ths expression equal to zero forces

$$\left(x + \frac{b}{2a}\right)^2 = \frac{b^2 - 4ac}{4a^2}$$

and a square root yields equation (5.4). This general solution for quadratic equations has been known for more than 1,000 years.

Finding roots to the general cubic equation $ax^3 + bx^2 + cx + d = 0$ is much more difficult; only in the early sixteenth century was this problem cracked. Credit goes to the Italians del Ferro, Tartaglia, and Cardano. The algebraic form of the solutions, though not impossible to write down, is daunting enough that the details are omitted here. Unlike the quadratic case, few mathematicians can reconstruct these equations on the spot or dictate the solutions by memory. The quartic (degree four) equation was solved essentially at the same time by another Italian, Ferrari. His solution, however, depended on solving the cubic, which he only learned about later, thus delaying the publication of his work. In the meantime, his mentor Cardano published solutions to both the cubic and the quartic in 1545.

Despite all this success, no one could solve the general quintic (degree five) equation. Notwithstanding their best efforts, the Italians could not budge this problem. Since mathematicians needed to find roots of polynomial equations, sometimes involving degrees that were are at least five, another route was taken. Researchers started developing numerical techniques that yielded accurate approximations to roots.

FIGURE 5.2: Niels Abel (left) and Évariste Galois (right).

Ironically, it is routine today to use these methods to approximate roots, even in the case of cubic and quartic polynomials.

After more than 250 years, the issue of quintic equations was finally settled. Significant progress was made by Paolo Ruffini in 1799, but a complete solution was only developed by the Norwegian Niels Henrik Abel (figure 5.2) in 1823. Presumably, many mathematicians expected that the solution to the quintic problem would be a more complicated cousin of the cubic and quartic formulas, but something more shocking was revealed. The Abel–Ruffini Theorem states that there is no general algebraic solution to polynomial equations of degree five or higher. This theorem does *not* say that every polynomial of degree five is not solvable; the equation $(x-1)(x-2)(x-3)(x-4)(x-5)=0$ has solutions $x=1,2,3,4,5$. The theorem says that there exist degree five polynomials whose roots cannot be written in terms of sums, differences, products, quotients, and radicals (as in the cubic and quartic cases). The equation $x^5-x+1=0$ is an example of an unsolvable equation.

Abel wrote up his solution in 1824 and disseminated it to several mathematicians. To save on printing costs, his note was written concisely in six pages. A longer, clearer version appeared later in a new journal established by August Leopold Crelle. Not only was Abel dogged by financial problems, but his health was deteriorating as well; he had contracted tuberculosis during a research visit to Paris. After hearing of Abel's health problems, Crelle used his influence to secure an appointment for Abel in Berlin. Unfortunately, it was too late. Abel passed away at the age of 26.

Shortly after, another young man, Évariste Galois of France, independently made a fundamental contribution to this area. Galois (pronounced gal-WA) found a way to distinguish whether a polynomial equation of degree five or higher could be solved. The work of both Abel and Galois was based on the then-new area of abstract algebra. Galois's methods are distinct enough that there is a branch of abstract algebra now referred to as Galois theory. Unfortunately, he did not live to see his mathematical ideas embraced by others. Besides his mathematical interests, Galois was a political firebrand who attracted trouble. He endured a six-month prison sentence for illegally wearing a uniform of the disbanded artillery of the National Guard. His troubles climaxed when he found himself in a duel. It is uncertain whether the duel was due to politics or over a woman. In any event, legend has it that Galois spent the night before the duel getting his mathematical discoveries down on paper. His concerns proved true; at age 20, Galois died the next day in the duel.

Diophantine Approximation

We encountered the Fibonacci numbers and the Golden Ratio back in chapter 1. Recall that the Fibonacci numbers are defined by $F_1 = F_2 = 1$ and

$$F_n = F_{n-1} + F_{n-2} \tag{5.5}$$

and the Golden Ratio ϕ is

$$\lim_{n\to\infty} \frac{F_n}{F_{n-1}} = \frac{1+\sqrt{5}}{2}$$

How well does F_n/F_{n-1} approximate ϕ? Diophantine analysis studies how well rational numbers can be used to approximate irrational numbers. Dirichlet's approximation theorem says that if α is any irrational number, then the inequality

$$\left|\alpha - \frac{p}{q}\right| < \frac{1}{q^2}$$

is satisfied by infinitely many integers p and q. If the right side was tightened, say to $0.5/q^2$, there would still be infinitely many solutions. Can we lower the constant 0.5 to a smaller number and still guarantee that infinitely many solutions exist? That is the substance of Hurwitz's Theorem: there are infinitely many solutions to the equation

$$\left|\alpha - \frac{p}{q}\right| < \frac{1}{\sqrt{5}q^2} \tag{5.6}$$

Moreover, if the constant $1/\sqrt{5}$ is made any smaller, the theorem breaks down. This is demonstrated by choosing α to be the Golden Ratio. Specifically, the equation

$$\left|\phi - \frac{p}{q}\right| < \frac{c}{q^2} \tag{5.7}$$

has at most a finite number of solutions if $c < 1/\sqrt{5}$. This gives yet another reason for Fibonacci fans to exchange high fives.

The Petersen Graph

Software verification usually means running a battery of tests and observing the software's response. Sometimes a well-designed test case can uncover easily missed defects. In mathematics, this philosophy is sometimes applied to calibrate a developing theory. In graph theory, the Petersen graph is an excellent "test graph" since it is a counterexample to many seemingly reasonable claims.

The Petersen graph is an undirected graph, usually depicted as a pentagon connected to an interior pentagram; see figure 5.3. The Petersen graph has a Hamiltonian path—a path that visits each vertex exactly once—but it does not have a Hamiltonian cycle; the

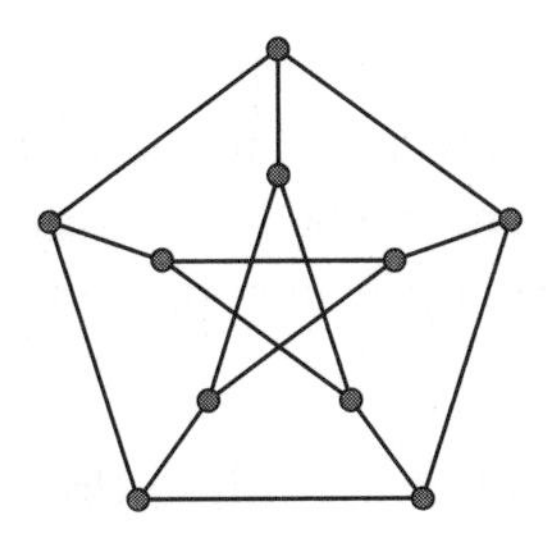

FIGURE 5.3: The Petersen graph.

ending vertex of a Hamiltonian path is never the starting vertex. However, the Petersen graph is the smallest example of a graph that is *hypohamiltonian*: it does not itself have a Hamiltonian cycle, but every graph formed by removing a single vertex *does* have a Hamiltonian cycle. Hypohamiltonian graphs arise in integer programming solutions to the Traveling Salesman Problem (TSP). The TSP, which asks for the shortest path needed to visit a set of prescribed cities, has widespread applications in resource allocation.

The Petersen graph is also the smallest example of a *snark*. A snark is a graph with the following properties:

- It is *connected*: there is a path from each vertex to every other vertex.
- It is *bridgeless*: removing any edge does not leave the graph disconnected.
- It is a *cubic graph*: each vertex supports three edges.
- Its *chromatic number* is 4: the minimum number of colors needed to color the edges so that no vertex has two edges of the same color equals four.

While the Petersen graph has these interesting properties, it is not planar. This means that any rendering of this graph in the plane must have edges that cross at a nonvertex point. Snarks have been of interest since 1880, when it was proved that the Four Color Theorem is equivalent to the statement that no snark is planar. This object was named by the famous mathematical writer Martin Gardner after the mysterious and elusive object of Lewis Carroll's poem "The Hunting of the Snark." Graph theorist William Tutte conjectured that every snark has a minor which is a Petersen graph (removing and contracting enough edges and vertices will lead to the Petersen graph). A proof of this result—dubbed the Snark Theorem—was announced in 1999, but as of 2013, the details are largely unpublished.

The Happy Ending Problem

Like other creative types, mathematicians get lost in their craft, sometimes seeing hours fly by like clouds in a time-lapse video. Their sense of beauty in learning—and especially discovering—mathematical truth parallels what most people feel when they are listening to blissful music. In this sense, every hard-fought problem that is finally solved can be called a happy ending. But Paul Erdős had another happy end in mind.

In 1933, a group of students in Budapest gathered frequently to discuss stimulating mathematics. Esther Klein proposed the following problem to the group: given five points in the plane, show that four of the points must form a convex quadrilateral. Erdős and George Szekeres were also part of the group. After Klein explained her solution, Erdős and Szekeres pushed this exploration farther, publishing a paper in 1935 that is now considered foundational to the area of combinatorial geometry. Klein's original problem bore other fruit as well; it brought Klein and Szekeres together. The two married in 1937, and Erdős therefore christened Klein's conundrum the Happy Ending Problem.

The solution to the problem is straightforward. First, find the convex hull of the set. If the convex hull is a pentagon, this means that any four of the five points form a convex quadrilateral. If the convex hull is a quadrilateral, the four corners of the convex hull do the job. If the convex hull is a triangle, the two other points are in the triangle's interior. Now consider the line through these two points. At least two of the triangle's points are on one side of the line. These two points plus the two interior points form a convex quadrilateral. The three cases are illustrated in figure 5.4.

In 1935, Erdős and Szekeres generalized this result. They showed that for any $n \geq 3$, there exists a smallest integer $N(n)$ such that any set of $N(n)$ points in general position in the plane contains n points, which form the vertices of a convex n-gon. Szekeres also conjectured that

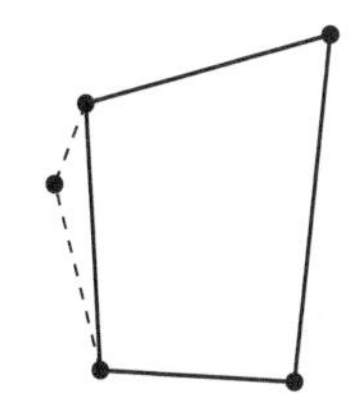

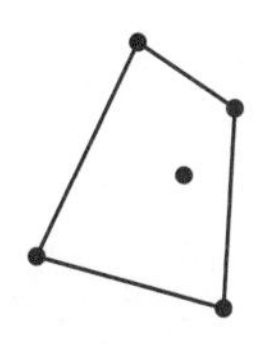

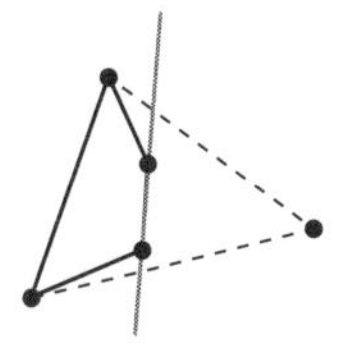

FIGURE 5.4: Identifying a convex quadrilateral when the convex hull has five points, four points, and three points.

$N(n) = 1 + 2^{n-2}$. Showing $N(3) = 3$ is obvious, and due to the Happy Ending Problem, we have $N(4) = 5$. Later it was shown that $N(5) = 9$. Values of $N(n)$ for $n \geq 6$, however, prove to be elusive. Shortly before Erdős died in 1996, he offered \$500 for a proof of Szekeres' conjecture.

Esther and George Szekeres faced political challenges soon after their marriage. Being Jewish, they left Europe in 1939 and lived challenging lives in Shanghai. In 1948, George accepted a position in Adelaide, Australia, where over the years his stature as a mathematician grew. George and Esther died within an hour of each other on August 28, 2005.

Tessellations

You have just been hired to tile a large room (don't worry; we'll give you advice). The owner, however, does *not* want you to use boring square or rectangular tiles. What other shapes allow you to tile the plane? Mathematicians call such arrangements *tessellations*.

If one uses only congruent, regular polygons, only three possibilities exist: squares, equilateral triangles, and regular hexagons. These are the *regular* tessellations. If more than one type of regular polygon is allowed and the set of the tiles around each vertex is identical, eight more arrangements exist. These are the *semiregular* or *Archimedean* tessellations. If you drop these restrictions, a myriad of possibilities exist.

Many artists and designers have extended the symmetry enjoyed in tessellations by adding patterns and drawings to the tiles. Beautiful tessellations can be found on walls and carpets in diverse cultures. The Alhambra palace in Spain is a showcase of tessellations in Islamic art. This artistry was a strong inspiration for the mesmerizing tessellations of Dutch artist M. C. Escher.

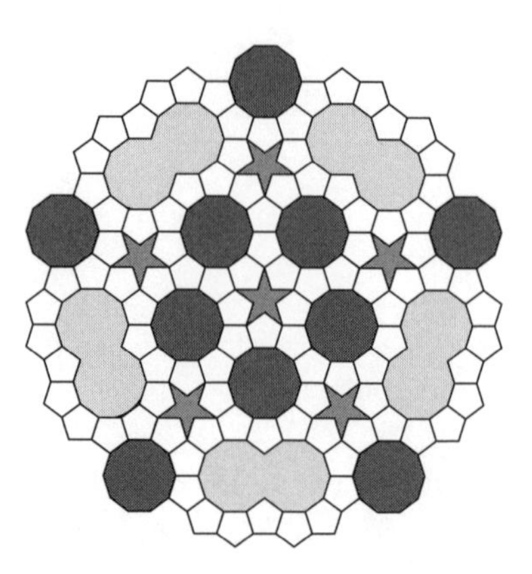

FIGURE 5.5: A tessellation combining pentagons, pentagrams, and fused decagons.

No tessellation can be made with only regular pentagons since the interior angle of 108° won't piece together (if you are a doubter, give it a shot). This didn't stop exploration into tessellations involving pentagons and other shapes. In the early sixteenth century, German artist Albrecht Dürer found a tessellation that uses regular pentagons and rhombi. A century later, Kepler—yes, the planets guy— discovered a tiling (figure 5.5) that uses regular pentagons, pentagrams, and "fused decagons."

Since regular tessellations tile the plane so easily, why bother going through contortions to tessellate with pentagons? Kepler's work inspired Roger Penrose to develop new tilings based on pentagons that tile the plane nonperiodically. This means that unlike the tessellations considered earlier, a nonperiodic tiling is not a shifted version of itself; it does not repeat. Such a tiling was considered to be nonexistent, but in the 1960s, a nonperiodic tiling with roughly 100 different tiles was constructed. Researchers successively found nonperiodic tilings that required fewer and fewer distinct tiles. Astonishingly, Penrose was able to reduce the number to two. The two different tiles are referred to as "kites" and "darts"; see figure 5.6. All the angles used in these tiles are integer multiples of $\pi/5$. So the regular pentagon, which resists being used alone for a tesllation, inspired the discovery of nonperiodic tilings.

How are the kites and darts pieced together to tile the plane? One way is to color the vertices black and white and require that adjacent

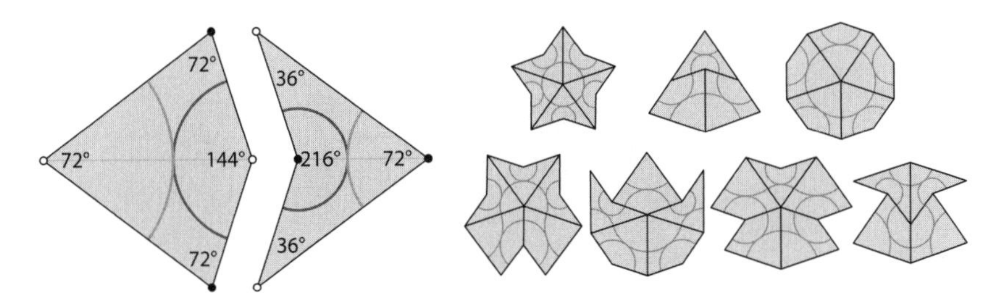

FIGURE 5.6: Kites and darts, and how they fit together.

tiles have matching vertices. Another matching procedure is to match the circular arcs in each tile.

The Penrose tiling is associated with the number 5 in another mysterious way. By slicing a five-dimensional hypercube with an "irrational" two-dimensional plane—a plane that misses any higher dimensional lattice—a Penrose tiling can be generated.

Nonperiodic tilings have also turned the field of crystallography upside down. Until the 1980s, the prevailing view of the structure of matter was that it contained only regularly repeating patterns. This traditional perspective claims that crystals possess only two-, three-, four-, and sixfold rotational symmetries. After a decade-long search, however, a quasiperiodic, mineralogical crystal—or quasicrystal for short—was found in 2009. Dubbed icosahedrite, this substance shows fivefold symmetry forbidden in the standard models. This quasicrystal is an alloy of aluminum, copper, and iron and occurs as tiny grains. Reportedly found in the Koryak Mountains of Russia, icosahedrite shows evidence of being extraterrestrial in origin. It is speculated that it was brought to Earth by an asteroid 4.5 billion years ago.

Of Balls and Sausages

When a nonmathematician uses the word "sphere," he or she usually means a ball or a globe. Mathematicians are usually more precise—some would say fussy or picky— in their use of language because this concept arises in many contexts. A *ball* is the set of points whose distance from a given center is less than or equal to the radius. A *sphere* is the set of points whose distance from the center is exactly equal to

Table 5.2
V_n, the Volume of the Unit n-Sphere

n	V_n	*Approximation*
1	2	2.0000
2	π	3.1415
3	$4\pi/3$	4.1888
4	$\pi^2/2$	4.9348
5	$8\pi^2/15$	5.2638
6	$\pi^3/6$	5.1677
7	$16\pi^3/105$	4.7248

the radius. In plain language, a ball is the whole solid, while a sphere is simply the thin shell, the surface of the ball.

We also need to let go of thinking of balls and spheres as objects only in three dimensions. Again, to be precise with language, we define the n-ball as the ball of radius 1 in dimension n and the n-sphere as the sphere of radius 1 in dimension $n+1$. The reason for this apparent discrepancy is that we want the volume of n-balls and the surface area of n-spheres to be measuring quantities in the same dimension. For example, the 1-ball is equivalent to the interval $[-1, 1]$, the set of all points that are at most one unit from the origin. The 1-sphere is a circle of radius 1. Note that both the 1-sphere and the 1-ball are one-dimensional objects. Similarly, the 2-ball is a disk—the filled-in circle—and the 2-sphere is the surface of a standard three-dimensional sphere of radius one. These objects are both two-dimensional. Let V_n denote the volume of the n-ball and S_n denote the surface area of the n-sphere. The first few values are $V_1 = 2$, $S_1 = 2\pi$, $V_2 = \pi$, and $S_2 = 4\pi$. There are tidy formulas that intricately link the volumes of balls and the surface area of spheres. In general, we have

$$V_{n+1} = \frac{S_n}{n+1}$$

$$S_{n+1} = 2\pi V_n$$

Focusing solely on volumes, these equations can be combined to form $V_n = 2\pi V_{n-2}/n$. Since $2\pi < 7$, we have the unintuitive claim that the maximum volume of an n-ball occurs when $n < 7$. Table 5.2 displays the first few values; the maximum of V_n occurs when $n = 5$.

There is another connection between higher dimensional balls and the number 5. Given an arrangement of n-balls in n-dimensional space, take the convex hull of this set. How can the balls be positioned so as to minimize the volume of their convex hull? Clearly we want to pack the balls tightly together, but the best configuration is not obvious. The Sausage Conjecture claims that when $n \geq 5$, the optimal configuration occurs when the balls are in a line so that the convex hull looks like a sausage. This quaint assertion has only been proved for large dimensions, specifically $n \geq 42$.

Knight's Tours on Rectangular Boards

In chapter 2, we saw that it is impossible to construct a knight's tour from one corner of a chess board to the opposite corner. The question as to whether *any* knight's tour is possible depends on the size of the board. For example, a 3×3 board certainly cannot have a knight's tour since the middle square can never be reached from any of the eight outer squares. If one side of the board is too short, a knight's tour is not possible. Conversely, a nicely packaged result claims that a knight's tour exists if both dimensions are at least 5 (figure 5.7).

There are several approaches to finding a knight's tour. From a programming perspective, a divide-and-conquer approach makes the most sense. This methodology slices the board into smaller pieces, constructs an appropriate tour on each piece, then assembles the parts together. An ad hoc approach known as Warnsdorff's rule asserts that the next move of the knight should be the square with the fewest number of forward moves. Some researchers have fine-tuned this approach in the case of ties, that is, when there are two or more squares with the same minimum number of forward moves. Lastly, whereas finding a knight's tour is a computational problem that can be tackled, the question of *how many* knight's tours exist for a given board is, in general, an unsolved problem.

Magic with Five Cards

The standard deck of 52 cards is a staple tool for magicians. You are about to learn an amazing card trick where the number 5 is used in

FIGURE 5.7: Knight's tour on a 24 × 24 board.

an obvious way. While this trick—known as the Fitch Cheney five-card trick—could be generalized (so in that sense, five is not special), it works so well with the familiar deck that it is worth mentioning.

But first a demonstration. The magician hands a deck of cards to a random volunteer for inspection and shuffling. The volunteer is then asked to choose five cards and hand them to the magician's lovely assistant. She passes four of the cards to the magician, one at a time, who lays them on a table: K♣, 7♢, 8♠, J♡. The magician announces that the fifth card, which the assistant has kept hidden, is 2♣. A gentle smile curls the assistant's mouth as she reveals the hidden card.

Assuming that the magician has convinced you that the cards have been chosen randomly and not marked, how does he divine the hidden card? The key involves the assistant. She has chosen which card to keep and the order in which to hand the other four to the magician. This provides enough information to tell the magician exactly what the hidden card is. Let's develop a system to do this. As we will see, the standard 52-card deck is perfectly tailored for this enthralling trick.

Since the assistant has five cards and there are four suits, there are at least two cards of the same suit. This is an instance of what mathematicians call the Pigeonhole Principle. If there are m pigeons who wish to rest in n pigeonholes, then one hole must contain at least $\lceil m/n \rceil$ pigeons, where $\lceil x \rceil$ denotes the smallest integer that is at least as large as x. For our application, there are five cards (pigeons) and four suits (holes), and since $\lceil 5/4 \rceil = 2$, two cards must have the same suit.

Next, we want to determine the "distance" between the two cards of the same suit (if there are more than two of the same suit, just choose any two of them). Identifying Jacks with the number 11, Queens with 12, Kings with 13, and Aces with 1, the distance between two cards is the minimal distance modulo 13. For example, the distance between a 9 and a Jack is 2. What about an Ace and a 10? Not 9, but 4 since the path 10–Jack–Queen–King–Ace requires only four steps. Since each suit has 13 cards, the distance between any two cards is at most 6.

From the two cards of the same suit, the assistant keeps the "higher" of the two cards. The "lower" card is the first card passed to the magician. This means that after the magician's first card is revealed, he knows the suit of the hidden card and has narrowed it down to six possibilities. The remaining three cards he will receive must do the work of identifying the hidden card exactly. To do this, we need a system that orders all cards in a deck. Using bridge ordering for suits—♣♢♡♠, which corresponds to alphabetical order—two cards can be ordered first by suit, and if they are the same, then by number. For example, K♣ $<$ 2♢. If the remaining three cards are labeled A, B, and C, representing the lowest, middle, and highest cards, then there are six possible orderings:

$$\{A, B, C\}, \{A, C, B\}, \{B, A, C\}, \{B, C, A\}, \{C, A, B\}, \{C, B, A\}$$

Interpret these lexigraphical orderings as the numbers 1 to 6. If the assistant hands the last three cards to the magician in the order so that the corresponding number equals the distance between the first card and the hidden card, the magician can calculate the mystery card exactly.

Let's see this algorithm in practice by applying it to the set of cards given at the beginning of this section: 2♣, K♣, 7♢, J♡, 8♠. Since there are two clubs and K♣ is two lower than 2♣, the hidden card is chosen to be 2♣. The first card passed to the magician is K♣. Since the distance is two, the assistant passes the cards in the order lowest, highest, and middle, which translates to 7♢, 8♠, and J♡.

Soccer Balls and Domes

As we saw earlier in this chapter, the dodecahedron is a Platonic solid whose 12 faces are regular pentagons. This solid can be modified to form other polyhedra that also have 12 pentgonal faces plus a bunch of hexagonal faces. Imagine "lifting" each face of the dodecahedron radially outward from the center of the solid. The size of the gap between faces can be chosen so that around each pentagon we place a ring of hexagons. This would add twenty hexagons and form the most beloved Archimedean solid in the world: the soccer ball (figure 5.8).

This process can be continued. Instead of adding a single ring of hexagons around each pentagon, choose a larger gap so that multiple layers of hexagons are added. No matter how many layers are added, there are always exactly 12 pentagons. Geodesic domes share this structure. The largest example is the dome used for Expo '67 in Montreal; see figure 5.8. Careful, though; you could hurt yourself looking for pentagons!

Recycling ad Infinitum

Recall that the Fibonacci numbers are a sequence of numbers where each number is built from the previous two. Let's try a different recurrence relationship where each term is generated from

$$x_n = \frac{x_{n-1} + 1}{x_{n-2}} \tag{5.8}$$

FIGURE 5.8: A soccer ball (top) and Montreal's geodesic dome at Expo '67 (bottom).

Table 5.3
Five Globally Periodic Recurrence Relationships

Recurrence	*Periodicity*
$x_n = x_{n-1}$	1
$x_n = 1/x_{n-1}$	2
$x_n = (x_{n-1} + 1)/x_{n-2}$	5
$x_n = x_{n-1}/x_{n-2}$	6
$x_n = (x_{n-1} + x_{n-2} + 1)/x_{n-3}$	8

If $x_1 = 4$ and $x_2 = 7$, then we find that $x_3 = 2$, $x_4 = 3/7$, $x_5 = 5/7$, $x_6 = 4$, and $x_7 = 7$. Since $x_6 = x_1$ and $x_7 = x_2$, this is enough to tell us that the five values will keep cycling forever. What if x_1 and x_2 are chosen differently? Amazingly, the values produced will almost always cycle with period five. To see this, let $x_1 = a$ and $x_2 = b$. The sequence of numbers then proceeds as

$$a, b, \frac{b+1}{a}, \frac{a+b+1}{ab}, \frac{a+1}{b}, a, b, \ldots$$

The only values of a and b that don't work are when one of the numbers in the sequence equals zero. This situation occurs when either $a = 0$, $b = 0$, $a = -1$, $b = -1$, or $a + b = -1$. The recurrence relationship (equation (5.8)) is sometimes called the *Lyness mapping*.

Researchers have looked for other globally periodic phenomena where the form of the recurrence relationship is also rational:

$$x_n = \frac{A_1 x_{n-1} + A_2 x_{n-2} + A_3 x_{n-3} + \cdots + A_{n-1} x_1 + A_n}{B_1 x_{n-1} + B_2 x_{n-2} + B_3 x_{n-3} + \cdots + B_{n-1} x_1 + B_n} \qquad (5.9)$$

where $A_1, \cdots, A_n$ and $B_1, \cdots, B_n$ are constants. All known recurrence relationships (5.9) that exhibit global cycling simplify to one of five possibilities. These are listed in table 5.3.

The Rogers–Ramanujan Identities

The Rogers–Ramanujan identities, first discovered by L. J. Rogers in 1894 and independently discovered by Srinivasa Ramanujan some time before 1913, are beautiful formulas connected to continued fractions

FIGURE 5.9: Srinivasa Ramanujan.

and the theory of partitions:

$$1+\sum_{n=1}^{\infty}\frac{q^{n^2}}{(1-q)(1-q^2)\cdots(1-q^n)}=\prod_{n=1}^{\infty}\frac{1}{(1-q^{5n-1})(1-q^{5n-4})} \tag{5.10}$$

$$1+\sum_{n=1}^{\infty}\frac{q^{n(n+1)}}{(1-q)(1-q^2)\cdots(1-q^n)}=\prod_{n=1}^{\infty}\frac{1}{(1-q^{5n-2})(1-q^{5n-3})} \tag{5.11}$$

In this chapter, we have already encountered the romantic stories of two mathematical geniuses who died young. It is therefore fitting to add another story, that of Ramanujan (1889–1920) (figure 5.9). Born to a poor Brahmin family in southern India, Ramanujan was attracted to mathematics as a youngster. He obtained a college scholarship based on his mathematical promise but was kicked out twice because of his indiscipline in studying nonmathematical subjects (in his case, English, Greek and Roman history, and physiology). Although Ramanujan's family was poor, they let him continue his mathematical investigations for the next five years without pressing him to seek employment. Ramanujan would spend his days working hard on his slate, discovering new equations involving infinite series, integrals, continued fractions, and special functions. When he found what he deemed a worthy result, he recorded it in his notebook. He carried his notebook everywhere and never lent it to anyone, fearing that the Indians would not value it and the English would steal it.

Though he was poor and had no degree, being a Brahmin afforded Ramanujan some connections. He doggedly sought exposure of his work in the growing Indian mathematical community. Eventually, he had a series of favorable hearings from Ramachandra Rao, who decided to support Ramanujan with a 25-rupee per month allowance. It was not much, but enough to free Ramanujan of financial worries.

Ramanujan was eventually encouraged to connect with prominent mathematicians in England. The first two ignored his letters, but the third, the Cambridge mathematician G. H. Hardy, had a different reaction. When Hardy and his close collaborator J. E. Littlewood read over the sampling of Ramanujan's mathematical results, they were overwhelmed. Some of the results were well known, having been established decades earlier; others required a little work to prove, and others seemed simply impossible to fathom. One of the many claims that captured Hardy's attention was a corollary of the Rogers–Ramanujan identities involving continued fractions:

$$\frac{1}{1+\frac{e^{-2\pi}}{1+\frac{e^{-4\pi}}{1+\frac{e^{-6\pi}}{1+\cdots}}}} = \left(\sqrt{\frac{5+\sqrt{5}}{2}} - \frac{\sqrt{5}+1}{2}\right) e^{2\pi/5}.$$

Hardy would write, "These equations had to be true because if they weren't, nobody could have imagined them" (Kanigel, *The Man Who Knew Infinity*, p. 111). Overcoming religious scruples, Ramanujan left India and collaborated with Hardy for the next five years. At one point, Ramanujan fell ill and was confined to a sanatorium. When he was eventually well enough to leave, he returned to India, supposing that the warmer climate and reconnection with his family would bring him back to full health. Upon his return, Ramanujan was welcomed as a hero and offered a professorship at the University of Madras. His health, however, did not improve, and he wasted away, dying at the age of 32.

The Rogers–Ramanujan identities, which intimately embrace the number five, concern the theory of partitions. The simplest partition function $p(n)$ counts the number of unordered ways one may write n as the sum of positive integers. For example, $p(4) = 5$ since there are five different ways to write the number 4 as a sum:

$$4 = 3 + 1 = 2 + 2 = 2 + 1 + 1 = 1 + 1 + 1 + 1$$

The partition function grows very quickly. A significant result due to Hardy and Ramanujan was the asymptotic formula

$$p(n) \approx \frac{1}{4\sqrt{3}n} e^{\pi\sqrt{2n/3}}$$

for large values of n.

The Rogers–Ramanujan identities imply two theorems for restricted forms of partitions. The first result claims that the number of partitions of n such that all parts differ by at least 2 is equal to the number of partitions of n such that all parts are congruent to 1 or 4 modulo 5. For example, if we take $n = 9$, then out of the 30 unrestricted partitions, five are of the first type,

$$9 = 8 + 1 = 7 + 2 = 6 + 3 = 5 + 3 + 1$$

and five are of the second type,

$$9 = 6 + 1 + 1 + 1 = 4 + 4 + 1 = 4 + 1 + 1 + 1 + 1 + 1$$

$$= 1 + 1 + 1 + 1 + 1 + 1 + 1 + 1 + 1$$

The second result is more refined: the number of partitions of n with minimal difference 2 and minimal part 2 equals the number of

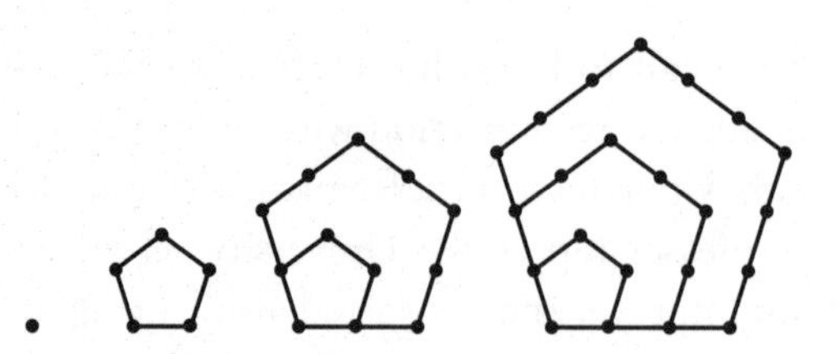

FIGURE 5.10: Pentagonal numbers.

partitions of n into parts congruent to 2 or 3 modulo 5. Again with $n = 9$, three are of the first type, $9 = 7 + 2 = 6 + 3$, and three are of the second type, $7 + 2 = 3 + 3 + 3 = 3 + 2 + 2 + 2$.

The number 5 relates to partitions in another important way. For this, we need to introduce the pentagonal numbers. We are familiar with squares and triangular numbers. The pentagonal numbers are constructed from pentagons that share a vertex and two sides (figure 5.10). The nth pentagonal number is the number of distinct dots contained within the pentagon which has n dots on each side. The first few are $1, 5, 12, 22, 35, 51, 70, 92, 117$, and if the g_n denotes the nth pentagonal number, then $g_n = n(3n - 1)/2$.

We can now present the head-turner known as the Pentagonal Number Theorem, due to Euler:

$$\prod_{n=1}^{\infty}(1 - x^n) = \sum_{k=-\infty}^{\infty}(-1)^k x^{g_k},$$

or in longhand,

$$(1 - x)(1 - x^2)(1 - x^3) \cdots = 1 - x - x^2 + x^5 + x^7 - x^{12} - x^{15} + \cdots$$

A basic result from the theory of partitions claims that

$$\prod_{n=1}^{\infty}\frac{1}{(1 - x^n)} = \sum_{n=0}^{\infty} p(n)x^n$$

These two equations can be combined to produce a recursive formula for the partition function:

$$p(n) = p(n-1) + p(n-2) - p(n-5) - p(n-7) + \cdots \quad (5.12)$$

For example, if we knew the values of $p(n)$ for all $n \leq 30$, then

$$p(31) = p(30) + p(29) - p(26) - p(24) + p(19) + p(16) - p(9) - p(5)$$

Using equation (5.12) recursively is a lightning-fast, easy-to-implement way to calculate the partition function exactly.

6

The Number Six

> Six is a number perfect in itself, and not because God created all things in six days; rather, the converse is true. God created all things in six days because the number is perfect. . . .
>
> — Saint Augustine, *The City of God*

The number 6 is technically a perfect number because all of its factors less than itself—1, 2, and 3—sum to itself. But 6 seems so perfect for structural or aesthetic reasons. As building honeycombs is beguiling to bees and stacking oranges mesmerizes grocers, the number 6 will cast a hex on you.

Optimal Packing

What do grocers and honeybees have in common? The obvious answer is that they are both adept at providing food for others. But there is a richer, more technical answer to this question: these two groups know how to efficiently pack their resources.

Honeycombs, made from the wax secreted by bees, are used to store honey, pollen, and larvae. For thousands of years, the honeycomb's hexagonal structure has been noted and admired (figure 6.1). It is wondered whether this entomological architecture inspired the interior

FIGURE 6.1: A honeycomb, nature's hexagonal tiling.

ribbing and hidden chambers in the dome of the Pantheon in Rome. Today, honeycomb structures have numerous engineering and scientific applications, including to the aerospace industry.

Why do honeycombs have a hexagonal structure? Pappus of Alexandria declared that bees "possessed a divine sense of symmetry," and Charles Darwin described the honeycomb as a masterpiece of engineering that is "absolutely perfect in economising labour and wax" (Peterson, "The Honeycomb Conjecture," p. 60). A mathematical rationale was given by the Polish polymath Jan Brożek (1585–1652): the hexagon tiles the plane with minimal boundary. Stated another way, Brożek conjectured that the optimal way to cover a large region with shapes of the same area while minimizing the boundary is to use the hexagonal structure. This problem resisted a solution for centuries but was finally positively settled by Thomas Hales in 1999.

Conveniently, the mathematical tools Hales used to prove the Honeycomb Conjecture were recycled after having settled another long-standing open problem, Kepler's Conjecture. Sometimes called the Cannonball Problem, it asks whether the traditional pattern used to

FIGURE 6.2: What's the best way to stack the oranges?

stack oranges (or cannonballs) is optimal (figure 6.2). Specifically, the arrangement of equal-sized spheres to fill a space with the minimum amount of wasted volume is the so-called close packing, which consists of layers of spheres whose centers form a hexagonal lattice. If equal-sized balls are dropped into a large container at random, experiments suggest that the density—the proportion of the container's volume filled by balls—is roughly 65%. The hexagonal close-packing arrangement yields an average density of $\pi/(3\sqrt{2}) \approx 74\%$. Efficient packing is the name of the game. This problem attracted the attention of mathematical giants such as Carl Gauss, who proved a special case. In 1900, David Hilbert's address at the International Congress of Mathematics in Paris described 10 problems that he felt would strongly influence the direction of mathematical research in the twentieth century. The print version of his address includes 23 problems. Kepler's Conjecture is part of Hilbert's Problem 18.

Like the Honeycomb Conjecture, the Kepler Conjecture languished for centuries until a solution was provided by Hales. In 1953, the

Hungarian mathematician László Fejes Tóth showed that the problem could be reduced to a finite (but very large) number of cases. He also grasped that such a challenge could be met with a computer, but this dream could not be realized at the time. Hales, assisted by his graduate student Samuel Ferguson, transformed the situation into minimizing a function of 150 variables. He showed that if the minimum of this function for 5,000 different configurations was larger than the minimum value obtained for the close packing, then the problem was solved. The solution involved a monstrous amount of computing, roughly 100,000 linear programming problems, an area of applied mathematics that is foundational to resource management. It took Hales several years to complete this project and settle the claim of Kepler's Conjecture.

As in the case of the Four Color Theorem, however, the solution attracted some controversy. The editors of the prestigious journal *Annals of Mathematics* agreed to publish it pending approval of a referee panel. After four years of work, the head of the panel, Gábor Fejes Tóth (son of László Fejes Tóth), reported that the panel was "99% certain" of the correctness of the proof but reserved claiming complete assurance because computer calculations could not be verified (Szpiro, "Does the Proof Stack Up?" pp. 12–13). While the mathematical community has generally accepted the proof, Hales has pursued a formal proof by using automated proof-checking software. In 2014, this project was announced to be complete.

Another packing problem is the Kelvin Conjecture. This is the three-dimensional version of the Honeycomb Conjecture, asking for the arrangement of three-dimensional cells of equal volume that minimizes the surface area between the cells. Lord Kelvin postulated that the solution consisted of filling the space with tetradecahedrons, a polyhedron with six square faces and eight hexagonal faces. Given the success of the Honeycomb and Kepler conjectures, one might be forgiven in quickly asserting that Kelvin's Conjecture must also hold. But wishful thinking is not sufficient. In 1993, Irish physicist Denis Weaire and his student Robert Phelan used computer simulations to find a more efficient cell that tiles three-dimensional space. This involves two different polyhedra with the same volume: an irregular dodecahedron with pentagonal faces, possessing tetrahedral symmetry,

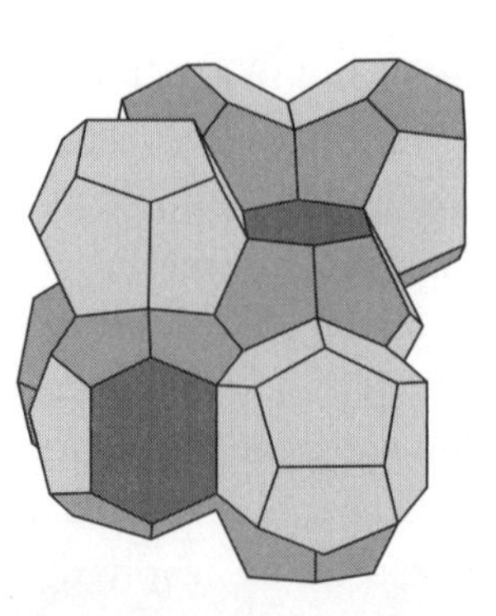

FIGURE 6.3: The Weaire–Phelan structure and Beijing's Water Cube aquatics center.

and a tetrakaidecahedron, which has two hexagonal and 12 pentagonal faces possessing antiprismatic symmetry (figure 6.3). This configuration has 0.3% less surface area than the arrangement asserted in the Kelvin structure, but it is unknown whether this is optimal.

The Weaire–Phelan structure is found in some crystal structures. Gas hydrates formed by methane, propane, and carbon dioxide at low temperatures have a structure in which water molecules lie at the nodes of the Weaire–Phelan structure. For a different application, this structure, like the honeycomb structure in two dimensions, is observed to be naturally strong. The Weaire–Phelan structure inspired the design of the Beijing National Aquatics Center—also known as the Water Cube—for the 2008 Olympics.

Of Friends and Strangers

Six people come together at a social gathering. If two of them have met before, we will call them friends, otherwise we say that they are strangers. The Friends and Strangers Theorem claims that either three people from the group are (pairwise) friends or three of them are (pairwise) strangers.

A proof of this theorem is short and sweet. Make a graph with six vertices, each representing a person, and an edge connecting each pair of vertices. If a pair of people are friends, make their connecting edge blue, otherwise make it red. Translating the statement of the theorem into the graph setting, our goal is to show that there must be a triangle whose edges are either all blue or all red.

Pick one of the vertices and call it P. Of the five edges emanating from it, at least three of them must be the same color (this is a simple instance of the Pigeonhole Principle). Label the vertices to which these three edges connect A, B, and C. Suppose the edges *PA*, *PB*, and *PC* are all red. If there was no red triangle in the graph, then none of the edges *AB*, *BC*, or *CA* could be red. This means, however, that they are all blue, and so there is a blue triangle. If the edges *PA*, *PB*, and *PC* were all blue, we would similarly have that *ABC* forms a red triangle. In all of these scenarios, we have proven the theorem.

Six Degrees of Separation

If two random people are chosen in the world, what is the shortest "friend chain" needed to connect them? A tall claim is that any person can be reached by anyone else with at most six steps, or six degrees of separation. This claim has been popularized by studies on various populations, including the work of psychologist Stanley Milgram in his article "The Small World Problem" from 1967.

While the idea of a universal bound on paths in networks gels with our increased communication abilities in a shrinking world, the actual number six is highly dependent on too many unrelated factors. Mathematically, it has been shown that if N represents the total number of nodes in a random network (think of people) and each node has K connections (think of friends), then the average path length between two nodes is $\ln N / \ln K$. Of course, determining K is difficult, thus making any concrete conclusions somewhat meaningless. Assuming that the world's population is 7 billion people, everyone would require 43 friends for the *average* path length to be six. Leaving the tenuous strength of social media relationships aside, this connection to six seems highly questionable.

As an aside, the advent of the computer after World War II spurred mathematicians to develop new algorithms that could be automated. Edsger Dijkstra's algorithm was established in the late 1950s to find the shortest path in a network. Such algorithms are the engine for technologies that provide the shortest path to one's destination.

So why include degrees of separation in this chapter if the connection to six is questionable? This gives us an excuse to ponder the connectivity in a population to certain people. In chapter 2, we encountered Paul Erdős, a commanding figure in twentieth century mathematics. Erdős led a singularly focused life where mathematics was pursued to the exclusion of all else. He made his living by collaborating with mathematicians in exchange for some financial support and being hosted. With no family or home, Erdős was the picture of the eccentric mathematician. His output, by any standard, was remarkable. If a mathematician retires, having published, say, 50 papers, this is considered very respectable. It's not hard to find some who have exceeded this amount. The prolific Erdős published more than 1,500 papers. Moreover, his network dwarfed everyone else's; he had more than 500 collaborators.

Imagine a graph where each node represents a mathematician and an edge between two nodes means that at least one paper was coauthored by those two mathematicians. This graph—called the collaboration graph—has been assembled and studied by Gerry Grossman at Oakland University. Based on data from 2004, the graph has roughly 400,000 nodes and about 2.9 million edges. The average number of authors per paper is 1.51, and the average number of papers per author is 7.21. Interesting trends can be witnessed over time, such as the growth in both papers authored and collaborators for the average mathematician. Before 1940, fewer than 10% of published papers had more than one author. Fifty years later, about 30% of papers were multiauthored.

Given the commanding production of Erdős, one's "distance" to him could be interesting. This inspires the notion of an *Erdős number*. If a

Table 6.1
Number of People with a Given Erdős Number

Erdős number	*1*	*2*	*3*	*4*	*5*	*6*	*7*	*8*	*9*	*10*	*11*	*12*	*13*
Number of people	504	6,593	33,605	83,642	87,760	40,014	11,591	3,146	819	244	68	23	5

person has coauthored a paper with Erdős, we say his or her Erdős number is 1. There are 504 people with an Erdős number of 1. If a mathematician's Erdős number is not 1, but he or she coauthored a paper with someone whose Erdős number equals 1, then his or her Erdős number is 2. This concept generalizes in the obvious way, basically measuring the distance between a given author and Erdős in the collaboration graph. Data from the Erdős Number Project's website list how many people have a given Erdős number; see table 6.1. Again, this information is based on data from 2004.

The median Erdős number is 5, and the mean is 4.65 (the author happily notes that his Erdős number is 3). The graph component connected to Erdős has 268,000 people. There are 84,000 isolated nodes in the graph (researchers who published only single-authored papers) and 50,000 who have collaborated but are not connected to Erdős. These people all have an infinite Erdős number.

A Necklace of Spheres

Take two mutually tangent spheres, S_1 and S_2, and surround them with another sphere S_3 to which they are internally tangent. We can always build a necklace of six spheres, each tangent to its two neighbors, and additionally tangent to S_1, S_2, S_3; see figure 6.4.

In fact, once the first sphere in the necklace is chosen, the rest of the spheres are uniquely determined. The centers of the six spheres all lie in a common plane, and their radii are related by

$$\frac{1}{r_1} + \frac{1}{r_4} = \frac{1}{r_2} + \frac{1}{r_5} = \frac{1}{r_3} + \frac{1}{r_6}$$

This fantastic claim is credited to Soddy (1937), but as with many mathematical results, it was observed much earlier. From the

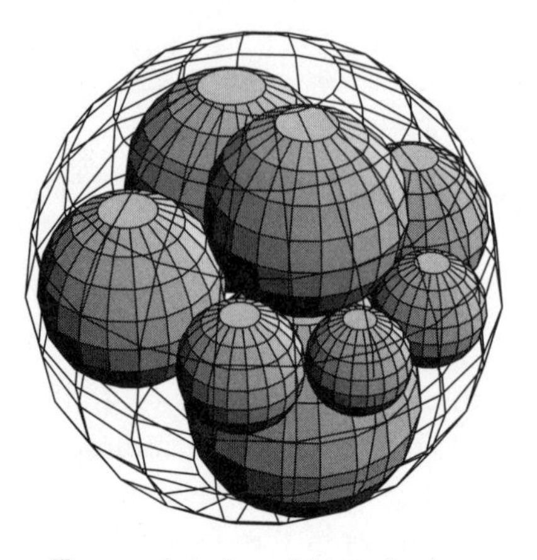
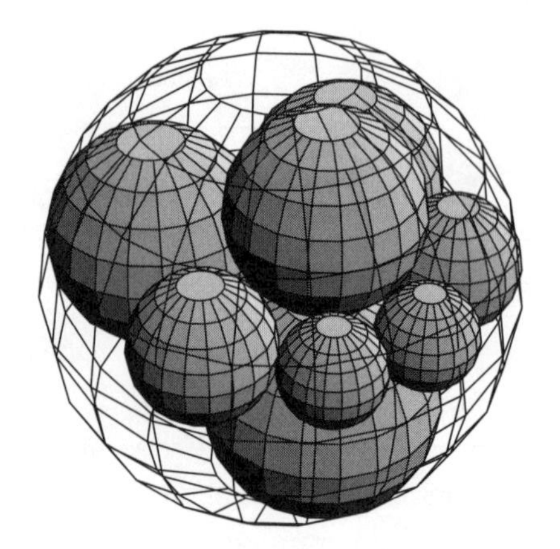

FIGURE 6.4: A necklace of spheres.

seventeeth to the nineteeth centuries, Japan witnessed a flourishing of mathematical problems—usually geometrical—called *sangaku*. These problems, meant for the enrichment of everyone, were written on wooden tablets and hung in Buddhist temples and Shinto shrines. The necklace of spheres problem was originally hung in 1822 by Yazawa Hiroatsu in the Samukawa shrine of Kozagun, Kanagawa prefecture.

Hexagons in Pascal's Triangle

With all the talk of hexagons in this chapter, it's almost novel to see hexagons make a nongeometric appearance. Starting with Pascal's triangle, pick any number that is in the interior, that is, not the number 1. Now take the product of the six numbers that surround it. For example, starting with either of the threes in the fourth row, we consider $1 \cdot 2 \cdot 3 \cdot 6 \cdot 4 \cdot 1 = 144$. Or choosing a four in the fifth row, we get $3 \cdot 1 \cdot 1 \cdot 5 \cdot 10 \cdot 6 = 900$. What do these products have in common? They are both perfect squares. Indeed, by choosing *any* interior number in Pascal's triangle, the product of its six surrounding numbers is always a square.

The proof is easy and is based on a finer structure. Split the set of six surrounding terms into two groups of three nonneighboring cells. For example, from figure 6.5, consider the two sets $\{6, 10, 35\}$ and $\{5, 20, 21\}$. The product of terms from each set is equal. To prove this in general, let the surrounded number be the binomial coefficient $\binom{n}{r}$.

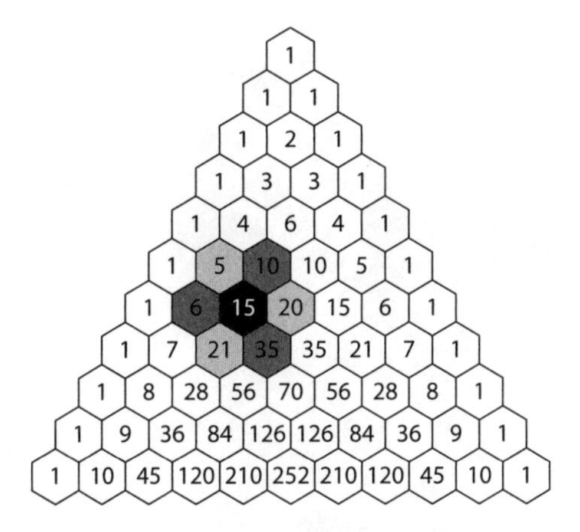

FIGURE 6.5: The product of the light gray cells equals the product of the medium gray cells.

Then we can compare the products of the two sets of three surrounding binomial coefficients:

$$\binom{n-1}{r-1}\binom{n}{r+1}\binom{n+1}{r}$$

$$= \frac{(n-1)!}{(r-1)!(n-r)!}\frac{n!}{(r+1)!(n-1-r)!}\frac{(n+1)!}{r!(n+1-r)!}$$

$$= \frac{(n-1)!}{r!(n-1-r)!}\frac{n!}{(r-1)!(n+1-r)!}\frac{(n+1)!}{(r+1)!(n-r)!}$$

$$= \binom{n-1}{r}\binom{n}{r-1}\binom{n+1}{r+1}$$

This implies that the product of the six surrounding terms is a square.

The Game of Hex

Tiling the plane with hexagons is so widespread that it has spawned a fascinating game: Hex. The game was first invented by the Danish mathematician Piet Hein in 1942. A few years later, it was independently reinvented by John Nash (who was the subject of the Oscar-winning film *A Beautiful Mind* and won a Nobel Prize in economics

FIGURE 6.6: The game of Hex.

in 1994). Early names for the game included "Polygon," "John," and "Nash," but Parker Brothers published the game in 1952 under the name "Hex," and this name has stuck.

Hex is a two-player game played on a rhombus-shaped board with hexagonal tiles (figure 6.6). The size of the board varies, but 11×11 is considered standard. Each player owns two opposite sides of the board. By alternately placing pieces on available tiles, the goal is to form a contiguous path from one of your edges to the opposite side.

A fascinating aspect of this game is that a tie is not possible; one player *must* win. It can be enlightening to experiment by filling the board with equal numbers of pieces in a random way and seeing that a path can always be formed. Better yet, try to *avoid* making a path for both players. At some point, a path will emerge, whether you like it or not. An interesting consequence of this no-tie property is that it can be used to prove the Brouwer Fixed-Point Theorem (from chapter 1) in two dimensions.

Since the player making the first move has an advantage, sometimes the so-called "pie rule" (or swap rule) is utilized. This says that after the first player makes his or her first move, the second player is given the option of trading places with the first player.

The Wendt Determinant

Mathematicians love to see surprising or nonintuitive results arise in their work. Such experiences raise the sense of wonder they experience while traversing a landscape of theorems. Matrix analysis, an area barely seen in this book, has some beautiful gems to show off.

A special class of $n \times n$ matrices are called *circulant*. Suppose that the first row of the matrix is specified. To construct the second row, shift all the entries of the first row one space to the right, with the last entry wrapping around to be the first. The third row is a shifted version of the second row, the fourth row is a shifted version of the third row, etc. If you've performed the shifts correctly, the first row will be a shifted version of the last row. Since the matrix is uniquely determined by the first row, we denote the matrix in terms of the n entries in the first row: $Circ(a_1, a_2, \ldots, a_n)$. Circulant matrices play a role in different areas of mathematics, including cryptography and graph theory. A particularly strong application is the discrete Fourier transform used in signal processing.

A connection was made in the late nineteenth century by E. Wendt between Fermat's Last Theorem and the determinant of a circulant matrix involving binomial coefficients:

$$W_n = \begin{vmatrix} 1 & \binom{n}{1} & \binom{n}{2} & \cdots & \binom{n}{n-1} \\ \binom{n}{n-1} & 1 & \binom{n}{1} & \cdots & \binom{n}{n-2} \\ \binom{n}{n-2} & \binom{n}{n-1} & 1 & \cdots & \binom{n}{n-3} \\ \vdots & \vdots & \vdots & \ddots & \vdots \\ \binom{n}{1} & \binom{n}{2} & \binom{n}{3} & \cdots & 1 \end{vmatrix}$$

The connection concerns the factors of W_n. Sometimes many prime factors arise, such as when $n = p - 1$ where p is an odd, prime number.

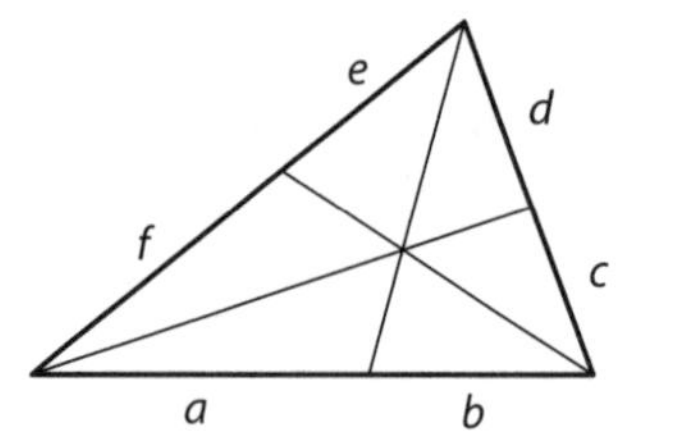

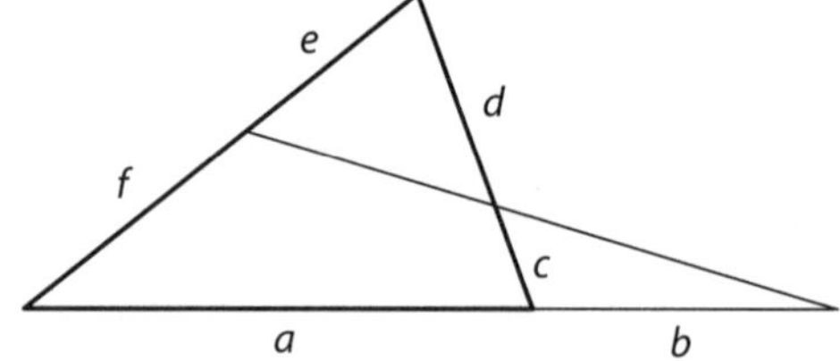

FIGURE 6.7: The theorems of Ceva and Menelaus.

In this case, p^{p-2} is a factor of W_{p-1}. Not surprisingly, the values of the Wendt determinant grow exceedingly fast, so numerical calculations become onerous in a hurry. Complete factorizations have been made for all values of $n \leq 500$. A curious pattern ties the Wendt determinant to this chapter: the number W_n equals zero if and only if n is a multiple of 6.

Six Lengths in Geometry

One should expect that there is less likelihood of having elegant relationships when more variables are thrown into a scenario. However, there are three elegant geometry theorems that involve six lengths. Sometimes it almost takes a sixth sense to see such connections materialize.

CEVA'S THEOREM

Pick any point within a triangle and draw the three lines that pass through this point and the three vertices, thus cutting each edge into two parts. The six quantities a, b, c, d, e, and f (figure 6.7) satisfy $ace = bdf$.

MENELAUS' THEOREM

This is similar to Ceva's Theorem, except that a line connects an edge of the triangle to an extended edge of the triangle. The equation is $(a + b)ce = bdf$.

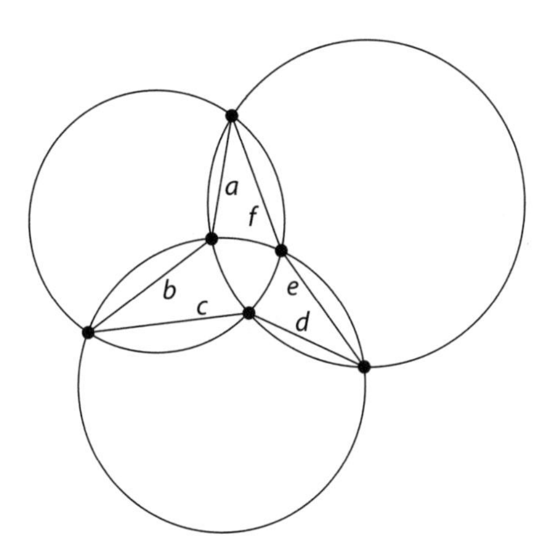

FIGURE 6.8: Haruki's Theorem.

HARUKI'S THEOREM

Instead of three triangles, Haruki considered the distances between various intersection points of three circles (figure 6.8). He found that $ace = bdf$.

7

The Number Seven

Isn't seven the most powerfully magical number?
—Tom Riddle, *Harry Potter and the Half-Blood Prince* by J. K. Rowling

If one is asked to pick a random number between 1 and 10, it seems that 7 is the most popular number. While an explanation is not apparent, seven surely has its charms and mystique. Even the digits of 1/7 point to some beautiful mathematics. We'll see how seven is special for multiplication, hearing the shape of a drum, and signal synchronization. And you won't need a sabbatical to get through it.

The Seven Circles Theorem

Although circles have already played a starring role in this book, the Seven Circles Theorem cannot be overlooked. This theorem is elementary—it requires no advanced mathematics to state or to prove—but it was only discovered in 1974. It makes you wonder how many other theorems requiring only low wattage are out there waiting to be discovered.

In any case, start with six circles forming a chain where each is tangent to its two neighbors, and all the circles are tangent to a seventh circle. From the first six circles, take pairs of opposite circles and draw a line connecting their points of tangency to the seventh circle. The Seven

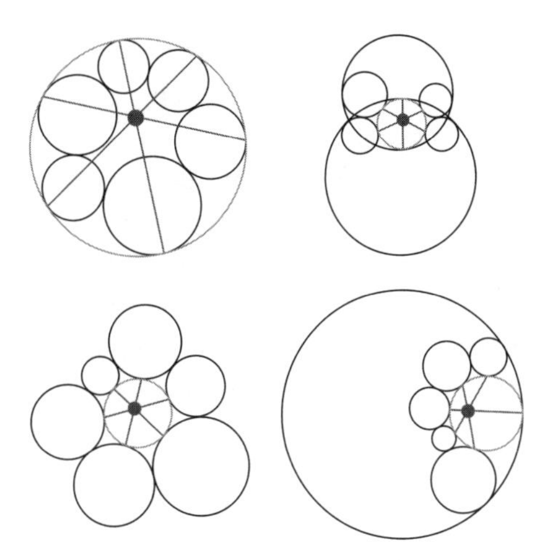

FIGURE 7.1: The Seven Circles Theorem.

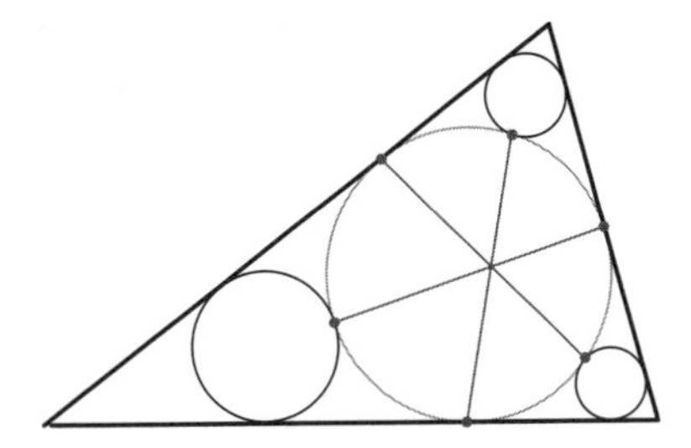

FIGURE 7.2: Circles in a triangle.

Circles Theorem says that the three lines intersect in a common point (figure 7.1). The theorem also works if the seventh circle is internally tangent, externally tangent, or a mixture.

An interesting, special case of the Seven Circles Theorem occurs when the six chained circles are exterior to the original circle. Let the radii of three alternating circles from the chain approach infinity (and the other three radii shrink accordingly). This converts these circles into lines and produces a result involving the inscribed circle of a triangle (figure 7.2).

Digits of 1/7 and Ellipses

The number seven is conspicuous among small numbers in that its reciprocal has an apparently messy representation: $1/7 = 0.\overline{142857}$.

For any integer $n \geq 2$, the decimal expansion of $1/n$ can have a period of at most $n - 1$. If this maximum period length is attained for a prime p, it is called a *long prime*, *golden prime*, or *maximal period prime*. The smallest long prime is seven. One might think that long primes are rare, but the opposite is the case. Even among small numbers, you don't have to sail the seven seas to find long primes; the first few are 7, 17, 19, 23, 29, and 47. In fact, long primes are suspected to be so plenteous that about 37.4% of primes are conjectured to be long. To be more precise, the exact proportion of primes that are expected to be long is given by Artin's Constant:

$$\prod_p \left[1 - \frac{1}{p(p-1)}\right]$$

where the product is taken over all primes p. Rather than recoil at such digital hogs, embrace the order in the digits. For starters, observe that

$$\frac{2}{7} = 0.\overline{285714}$$

$$\frac{3}{7} = 0.\overline{428571}$$

$$\frac{4}{7} = 0.\overline{571428}$$

$$\frac{5}{7} = 0.\overline{714285}$$

$$\frac{6}{7} = 0.\overline{857142}$$

Note that for all of these fractions, the order of the digits does not change.

More interesting, however, is that these digits can be used to form special ellipses. The equation for an ellipse takes the general form $Ax^2 + Bxy + Cy^2 + Dx + Ey + F = 0$. Since this equation can be scaled, this leaves five degrees of freedom, meaning that there is typically one ellipse going through a set of five points. The word "typically" is circumvented by an ellipse containing the six points $(1, 4)$,

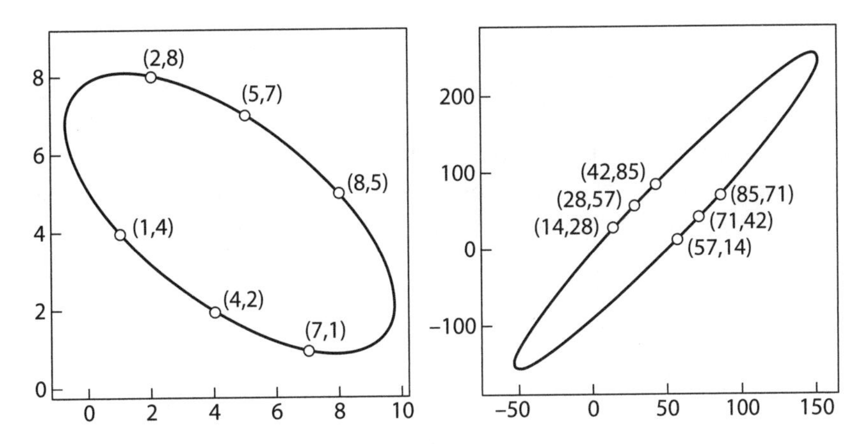

FIGURE 7.3: Two one-seventh ellipses.

(4, 2), (2, 8), (8, 5), (5, 7), and (7, 1), made from the repeating digits of 1/7. This has been called the *one-seventh ellipse*, and its equation is

$$19x^2 + 36xy + 41y^2 - 333x - 531y + 1,638 = 0$$

Even more eye-popping, another ellipse can be formed with the points (14, 28), (42, 85), (28, 57), (85, 71), (57, 14) and (71, 42) (figure 7.3). Its equation is

$$-165,104x^2 + 160,804xy - 41,651y^2 + 8,385,498x$$
$$-3,836,349y - 7,999,600 = 0$$

Strassen's Matrix Multiplication

Matrix multiplication is one of the most common calculations used in matrix algebra. For example, 2×2 rotation matrices are routinely used in computer graphics. Every student learns that if A and B are defined as

$$A = \begin{bmatrix} a_{11} & a_{12} \\ a_{21} & a_{22} \end{bmatrix}, \quad B = \begin{bmatrix} b_{11} & b_{12} \\ b_{21} & b_{22} \end{bmatrix}$$

then their product is

$$C = AB = \begin{bmatrix} a_{11}b_{11} + a_{12}b_{21} & a_{11}b_{12} + a_{12}b_{22} \\ a_{21}b_{11} + a_{22}b_{21} & a_{21}b_{12} + a_{22}b_{22} \end{bmatrix}$$

Note that to calculate AB, one requires 8 multiplications and 4 additions. Since multiplication requires much more computer memory than addition, it would be valuable to reduce the number of multiplications, even at the expense of having a few more additions. Such an improvement was effected by Volker Strassen in 1969 and is now called Strassen multiplication. Define seven new terms, each involving exactly one multiplication:

$$m_1 = (a_{11} + a_{22})(b_{11} + b_{22})$$

$$m_2 = (a_{21} + a_{22})b_{11}$$

$$m_3 = a_{11}(b_{12} - b_{22})$$

$$m_4 = a_{22}(b_{21} - b_{11})$$

$$m_5 = (a_{11} + a_{12})b_{22}$$

$$m_6 = (a_{21} - a_{11})(b_{11} + b_{12})$$

$$m_7 = (a_{12} - a_{22})(b_{21} + b_{22})$$

Then the terms for the matrix C can be calculated as

$$c_{11} = m_1 + m_4 - m_5 + m_7$$

$$c_{12} = m_3 + m_5$$

$$c_{21} = m_2 + m_4$$

$$c_{22} = m_1 - m_2 + m_3 + m_6$$

With these intermediate terms, one now calculates the product AB with 7 multiplications and 18 additions or subtractions.

Strassen multiplication is not limited to 2×2 matrices. If the two matrices have size $2^n \times 2^n$, then each matrix can be divided into four $2^{n-1} \times 2^{n-1}$ blocks. One can now apply Strassen multiplication described earlier, not with numbers, but with these four blocks. In fact, this process can be used recursively on each block to reduce the number of multiplications even more. This means that to multiply two $N \times N$ matrices, we can reduce the number of multiplications from roughly N^3 to $N^{\log_2 7} \approx N^{2.8}$.

With computer architectures where multiplication is only marginally more computationally expensive than addition, Strassen's algorithm only effects savings if the matrices are sufficiently large. Moreover, demands on computer memory are substantially higher for Strassen multiplication over the standard approach, so care must be taken because of this overhead.

The Fano Plane

Mention the word "geometry" and most people think of sets in the plane (lines, points, circles, rectangles, etc.) or three-dimensional structures. The physics-minded person may imagine four-dimensional space, where an extra dimension representing time has been appended. Mathematicians have extended the concepts of geometry in many ways. An apparently simple direction is *finite geometry*, a geometric scenario that contains only a finite number of points. That there are only finitely many pixels on a computer screen—or for that matter, finitely many particles in the universe—should convince you that this is not such a far-removed concept.

Let's restrict our attention to *projective spaces*. Bending the popular rules of geometry some more, "lines" in these spaces may not remain straight, but they do connect points together. The conditions we prescribe on these spaces are as follows:

1. For every pair of points, there is a unique line joining them,
2. Every pair of lines intersects in a unique point,
3. There exist four points, no three of which belong to the same line.

Note that the first condition violates our sense of lines being parallel. Of course, this concept has been used by artists for centuries with perspective drawing. The simplest example of a finite, projective space is the *Fano plane* (figure 7.4). It consists of seven points and seven "lines." There are three points on every line and three lines through every point (the circle is one of the "lines").

The seven points in the Fano plane can be represented by the seven nonzero, three-digit binary numbers. The locations of these numbers

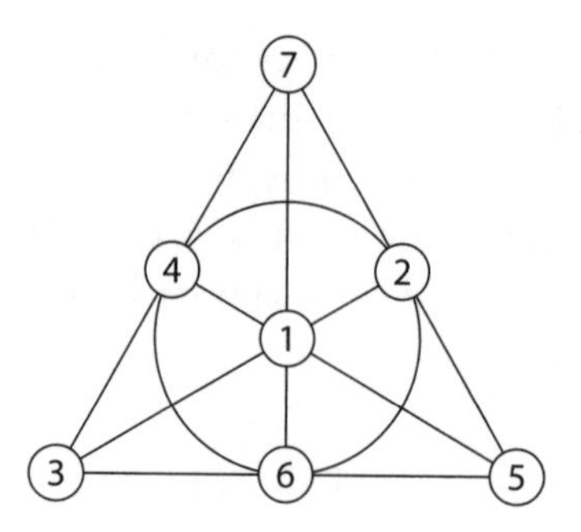

FIGURE 7.4: The Fano plane.

is not haphazard. By taking any two points, the third point on the common line can be obtained by adding the two numbers bitwise modulo 2. This can also be thought of as adding the numbers in base 2 but ignoring any carries. For example, take the numbers 3, 5, and 6, which are on one of the lines. By summing the binary digits of any two of these numbers and ignoring carries, we get the third number:

$$011 + 101 = 110$$

$$101 + 110 = 011$$

$$110 + 011 = 101$$

The lines of the Fano plane can also be identified with the seven nonzero, three-digit binary numbers. This identification is made by choosing the unique triple so that the dot-product of the line's number with each of the points on the line, modulo 2, equals zero. (The dot product of (a,b,c) with (d,e,f) is written $(a,b,c) \cdot (d,e,f)$ and equals $ad + be + cf$.) For example, the line containing 3, 4, and 7 is identified with the binary number 011 since

$$(0,1,1) \cdot (0,1,1) \equiv 0 \pmod{2}$$

$$(0,1,1) \cdot (1,0,0) \equiv 0 \pmod{2}$$

$$(0,1,1) \cdot (1,1,1) \equiv 0 \pmod{2}$$

An amusing application of the Fano plane concerns the so-called Transylvania lottery. For each ticket, a player picks three numbers from 1 to 14. When the draw is performed, three numbers from 1 to 14 are also chosen. A ticket wins if at least two of the numbers match. A basic question asks how many tickets—out of a possible $\binom{14}{3} = 364$—should a player buy to guarantee a win? The answer is 14. In fact, here are the ticket numbers that work:

1–2–3, 1–4–5, 1–6–7, 2–4–6, 2–5–7,

3–4–7, 3–5–6, 8–9–10, 8–11–12, 8–13–14,

9–11–13, 9–12–14, 10–11–14, 10–12–13.

It's not hard to check that every pair of numbers is represented on one of these 14 tickets. By the Pigeonhole Principle, either two of the numbers of the winning triple are low (between 1 and 7) or two numbers are high (between 8 and 14). The first seven tickets listed contain exactly one ticket with a given low pair, and the second seven contain exactly one ticket with a given high pair. Why does this solution work? Recall that each pair of points contained in the Fano plane has a line going through it. For the low range, we simply need to choose the triples corresponding to the seven lines. By adding seven to every number in the low tickets, we obtain the seven high tickets. These 14 triples cover each pair of numbers exactly once.

Border Patterns

After having painted the dining room, your partner asks about putting a border near the ceiling. This means a linear pattern that is periodic, that is, repeats itself after a fixed distance. After going through many patterns online—too many, you think—you notice that besides the symmetry of the repetition, some of the patterns have other symmetries as well. In fact, *all* the patterns can be classified into one of seven different possibilities. This collection is known as the *Frieze group*. Table 7.1 describes the seven general patterns.

Table 7.1
Frieze Patterns

Name	Description	Example
Hop	Only the translation.	
Sidle	Translations and reflections across some vertical lines.	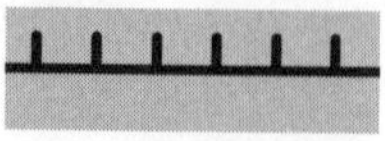
Jump	Translation and reflection across the horizontal axis. These two symmetries can be combined to exhibit glide reflections.	
Step	Translations and glide reflections. Can you see the difference between a jump and a step?	
Spinning hop	Translations and 180° rotations around some points on the horizontal axis.	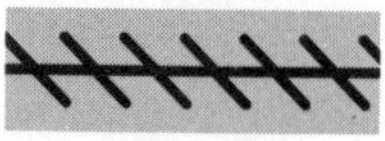
Spinning sidle	Besides the sidle's symmetries, there are also glide reflections and 180° rotations around some points on the horizontal axis.	
Spinning jump	This one has everything: translations, vertical and horizontal reflections, glide reflections, and 180° rotations.	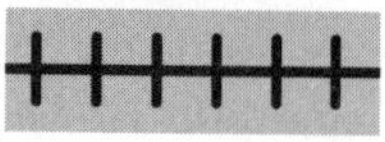

The Szilassi Polyhedron and the Heawood Graph

What do a sphere and a tetrahedron have in common? Obviously, a sphere is a smooth surface, whereas a tetrahedron is a polyhedron with four faces. From a topological perspective, however, the two solids are the same. The sphere—think of a ball of clay—can be sculpted into a polyhedron without breaking off parts or introducing holes.

Can we do the same with the torus? In other words, is there a polyhedron that has the same topology as the torus? There are plenty of possibilities, but which one has the fewest number of faces? In 1977, Lajos Szilassi discovered a polyhedron with seven faces that is topologically equivalent to the torus (figure 7.5). The *Szilassi polyhedron* has 7 hexagonal faces, 14 vertices, and 21 edges. Curiously, each face shares an edge with every other face. The tetrahedron is the only other polyhedron known to have this property.

The graph that represents this polyhedron is called the *Heawood graph*. To get the torus from the square, first wrap the right and

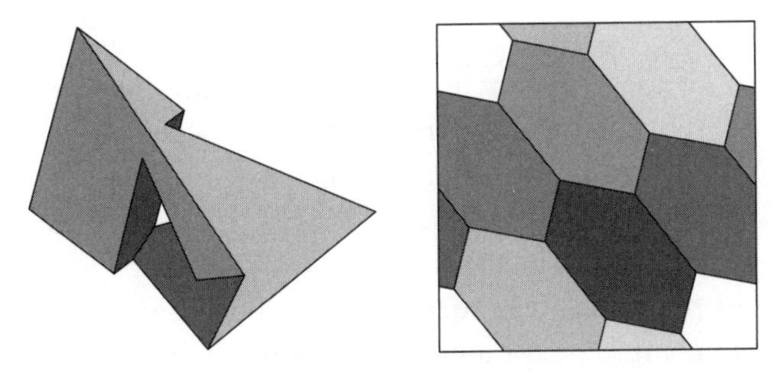

FIGURE 7.5: The Szilassi polyhedron and the Heawood graph.

left edges around to form a vertical tube (you should see that the shadings match up), then bend the tube to join the two circular edges. Of course, the graph shares the same number of vertices, faces, and edges as the polyhedron. When embedded on the surface of a torus, however, there are no crossings. The graph's discoverer, Percy John Heawood, proved in 1890 that in every subdivision of the torus into polygons, the polygonal regions can be colored by at most seven colors. Since the Heawood graph requires seven colors (figure 7.5), this shows that Heawood's theorem cannot be improved (mathematicians say this result is *sharp*). By taking the dual of the Heawood graph, one finds another polyhedron with the same topology as the torus. Named after another Hungarian, Ákos Császár, and discovered in 1949, the Császár polyhedron has 7 vertices, 21 edges, and 14 triangular faces.

In 1890, when researchers were far away from proving the Four Color Theorem, Heawood thought about graph coloring for graphs that could be imbedded on tori with g holes. He showed that the minimum number of colors needed to color all graphs on surfaces with g holes is

$$\left\lfloor \frac{7+\sqrt{1+48g}}{2} \right\rfloor$$

The torus is the special case with $g = 1$.

The Kuratowski Closure–Complement Theorem

Suppose that S is a subset of the real line. We want to see how many different sets we can construct by successively taking one of two operations: closure or complement. What do these mean? The *complement* of a set S—not to be confused with *compliment*, as in, "That's a lovely hat you're wearing, Madame Bellemare"—means all the points on the real line that are not in S.

The other concept, *closure*, is a bit trickier. It's best to start with an example. Let $S = [0, 1)$ be the set of points x such that $0 \leq x < 1$. The sequence of points 0.9, 0.99, 0.999, etc., all lie in S and get arbitrarily close to the point 1, even though $x = 1$ is not itself in S. This means that $x = 1$ is a *limit point* of the set S. We could go in the other direction and define the sequence of points 0.1, 0.01, 0.001, etc., to get arbitrarily close to $x = 0$, but since this point is already in S, it is not a limit point. So what does the closure of a set mean? It is the set S plus all of its limit points. Thus the closure of $[0, 1)$ is $[0, 1]$.

Now let's go back to the question of making new sets with closures and complements. It's not hard to see that the complement of the complement of S is simply S; applying repeated complements is like toggling back and forth between two windows on a screen. Less obvious, though believable, is that once we have taken the closure of S, taking the closure again adds no new points. In other words, once you've augmented a set by adding its limit points, trying to take the closure again adds nothing new. Using symbols, let c denote closure and k denote complement. We thus have that

$$kkS = S \tag{7.1}$$

and

$$ccS = cS \tag{7.2}$$

With these two identities, the only possible new sets that can be constructed with closures and complements involve alternately applying c and k to the set S. Of course, this could go on indefinitely, until we mention that there is another useful identity: $ckcS = ckckckcS$. This is even thornier to explain than closure. The *interior* of a set, denoted as iS, is (loosely speaking) the set S minus its boundary points. This can be formally defined as $iS = kckS$. For example, the interior of $[0, 1)$ is $(0, 1)$, where you see that we have removed the point

$x = 0$. Since the interior of $ckcS$ is contained in the set itself, we have $kckckcS \subseteq ckcS$. Taking the closure of each side and using $cc = c$, we obtain $ckckckcS \subseteq ckcS$. On the other hand, $kckcS \subseteq cS$ since $kckcS$ is the interior of cS. Then $ckckcS \subseteq cS$, $kckckcS \supseteq kcS$, and $ckckckcS \supseteq ckcS$. These combine to produce

$$ckckckcS = ckcS \tag{7.3}$$

Equations (7.1)–(7.3) show that there are at most 14 distinct possibilities that one can produce starting with the set S:

$$S, kS, ckS, kckS, ckckS, kckckS, ckckckS, kckckckS, cS, kcS,$$
$$ckcS, kckcS, ckckcS, kckckcS$$

Note that the maximum number of applications of c and k is seven. This result, proven in 1922, is known as the Kuratowski Closure–Complement Theorem.

There is one issue left to resolve. The preceding logic shows that there are *at most* 14 different sets created. Is there an example where all 14 sets are distinct? After all, some of these sets may be the same. For example, if we start with $S = [0, 1) \cup [2, 3)$, we find that there are only six possibilities:

$$S = [0,1) \cup [2,3)$$
$$cS = [0,1] \cup [2,3]$$
$$kcS = (-\infty,0) \cup (1,2) \cup (3,\infty)$$
$$ckcS = (-\infty,0] \cup [1,2] \cup [3,\infty)$$
$$kckcS = (0,1) \cup (2,3)$$
$$kS = (-\infty,0) \cup [1,2) \cup [3,\infty)$$

To get all 14 sets to be different, a fancier set S is needed. Can you find one? See Chapter 10 for a solution.

Can You Hear the Shape of a Drum?

How does a mother distinguish the voices of her children? A conductor the instruments of the orchestra? A prey the sounds of the jungle? The common answer is that each noise makes its own distinct sound. Or does it? In 1966, the mathematician Mark Kac wondered whether two different-shaped drum heads could be audibly indistinguishable.

A drum generates sound as the head vibrates. The vibrations can be decomposed into different components—these are called *modes*—each vibrating at their own frequency. This set of frequencies is called the *spectrum* of the drum. The modes and frequencies are modeled with a partial differential equation. If D is the shape of the drumhead, the modes of vibration are solutions $u = u(x, y)$ to the Helmholtz equation

$$\frac{\partial^2 u}{\partial x^2} + \frac{\partial^2 u}{\partial y^2} + \lambda u = 0 \qquad (7.4)$$

The frequencies are represented by the Greek letter λ, and the function $u(x, y)$ represents the height of the drumhead above its equilibrium position. Since the skin is clamped down at the perimeter of the drum, we have $u = 0$ there.

The simplest case to consider mathematically is when the region D is a square. In this special scenario, one can actually specify the modes explicitly with trigonometric functions (closed-form representations for more complicated shapes of drumheads are usually not possible). Assuming that the square has side length 1, some modes are given by the functions $\sin(m\pi x)\sin(n\pi y)$, where m and n are positive integers. Along the four boundary lines—$x = 0$, $x = 1$, $y = 0$, or $y = 1$—the modes equal zero. Using equation (7.4), one finds that the frequencies λ equal $\pi^2(m^2 + n^2)$. Figure 7.6 shows an example. Note that it is possible to have two different modes with the same frequency. For example, taking $(m, n) = (1, 8)$, $(8, 1)$, $(4, 7)$, and $(7, 4)$ all produce a frequency of $65\pi^2$. Modes with the same frequency can be combined to produce new modes (also with the same frequency).

In an attempt to visualize the different modes, German physicist Ernst Chladni (1756–1827), the "father of acoustics," did research on vibrating sheets of metal. Running a violin bow across the edge of a

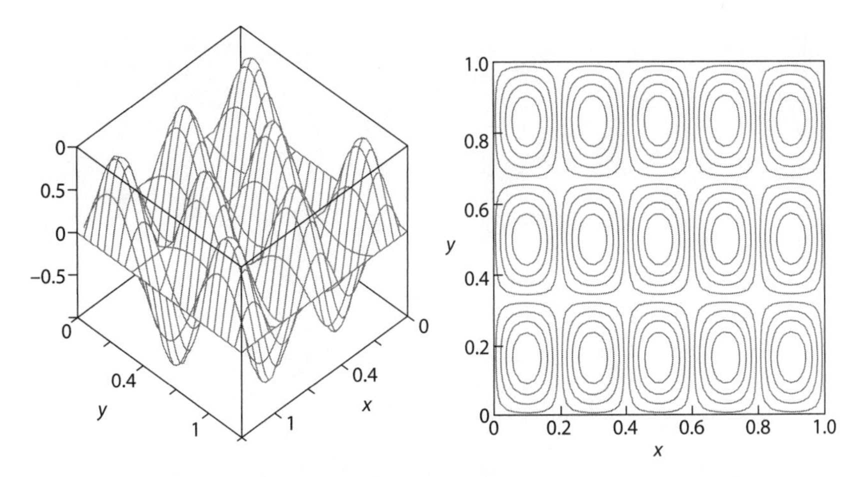

FIGURE 7.6: Modes of vibration and contour plots, $m = 5, n = 3$.

metal plate excites the plate into resonance. For most modes, there are places interior to the square where the height of the mode equals zero at all times. These collections of "dead points" form curves called *nodal lines*. Points off the nodal lines vibrate up and down. Chladni's idea was to sprinkle a thin layer of sand on the plate before setting it into motion. When the plate is excited and goes into resonance, the sand dances across the plate and migrates to the nodal lines since there is no motion there. This process reveals the beautiful nodal line patterns (figure 7.7).

Now back to Kac's question, which really asks, "Can two different regions have the same spectrum?" If they do, we say that the two regions are *isospectral*. A mathematical analysis reveals that isospectral regions must be identical in many respects, including having the same area and the same perimeter. Since a spectrum is infinite—there are infinitely many overtones—there are infinitely many commonalities between isospectral regions. It is not unreasonable to surmise that all these shared properties would prohibit two different domains from being isospectral. After all, if it looks like a duck, quacks like a duck, and flies like a duck, mustn't it be a duck? Surprisingly, the answer is no.

In 1992, Carolyn Gordon, David Webb, and Scott Wolpert showed that isospectral domains exist. Their proof uses advanced mathematical machinery and rests on Sunada's Theorem, which offers conditions

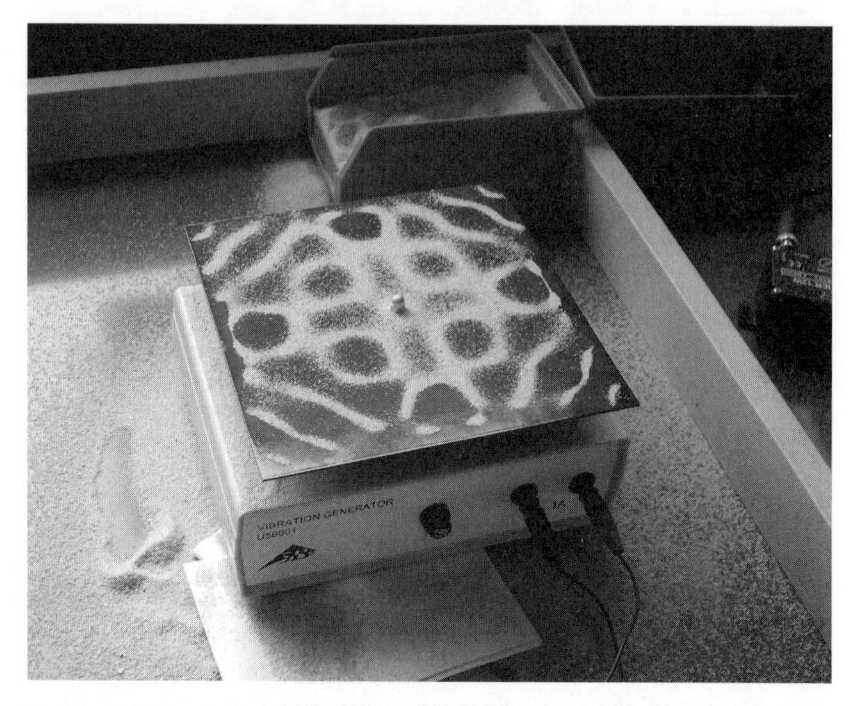

FIGURE 7.7: Sand patterns on vibrating plates.

guaranteeing that two domains are isospectral. The explicit example they gave is in figure 7.8. You'll note that each domain is a patchwork of seven congruent right triangles. This is no accident. Other researchers have explored this example further and have developed a relatively simple procedure that constructs the mode of one domain from the mode of the other. This approach works because the seven triangles are connected in a similar way to the seven points in the Fano plane. In fact, other isospectral domains can be constructed by identifying the connection to other projective spaces. In the end, the properties of some abstract algebraic structures allow one to answer questions about some seemingly far-removed problems in acoustics.

Barker Codes

Suppose that two devices transmit the same signal but are suspected of being out of synch. To check this, one would like a simple test that would show a marked difference between synchronized and

FIGURE 7.8: Modes of vibration for two isospectral shapes and decomposition into seven congruent triangles.

unsynchronized signals. To this end, one uses a *Barker code*, a finite sequence of $+1$ and -1 terms. If the code has n terms, then the autocorrelation of the code $\{a_j\}$ shifted by k is defined as

$$c_k = \sum_{j=1}^{n-k} a_j a_{j+k}$$

For example, for the code with $a_1 = 1$, $a_2 = 1$, $a_3 = -1$, and $a_4 = 1$, we have

$$c_0 = 1 \cdot 1 + 1 \cdot 1 + (-1) \cdot (-1) + 1 \cdot 1 = 4$$

$$c_1 = 1 \cdot 1 + 1 \cdot (-1) + (-1) \cdot 1 = -1$$

$$c_2 = 1 \cdot (-1) + 1 \cdot 1 = 0$$

$$c_3 = 1 \cdot 1 = 1$$

Note that the peak autocorrelation ($k = 0$) produces the value 4, but the off-peak autocorrelations ($k > 0$) produce -1, 0, and 1. This is an example of a Barker code: the peak autocorrelation equals n, and the

Table 7.2
Seven Known Barker Codes

n	*Barker Code*
2	$++$
3	$++-$
4	$+++-$
5	$+++-+$
7	$+++--+-$
11	$+++---+--+-$
13	$+++++--++-+-+$

off-peak autocorrelations lie in the set $\{-1, 0, 1\}$. Because of this drastic change in values between peak and off-peak autocorrelations, Barker codes give strong evidence that two signals are synchronized or not.

The Barker code from the example can be written as $\{++-+\}$. What other Barker codes can be constructed? Note that a Barker code can be transformed to produce other Barker codes. By manipulating the sum, one can show that if $\{a_k\}$ is a Barker code of length n, then so is $\{-a_k\}$, $\{a_{n-1-k}\}$, and $\{(-1)^k a_k\}$. After accounting for these knockoffs, there are only seven known Barker codes; (table 7.2).

It has been shown that all the possible Barker codes with odd n have been found, but it is unresolved whether there are any more with even n. If there is another Barker code, theoretical and numerical work has shown that $n > 2 \cdot 10^{30}$. Well, except for possibly one value: $n = 189{,}260{,}468{,}001{,}034{,}441{,}522{,}766{,}781{,}604$.

Recreational Mathematics

This book has resisted compiling facts about numbers that do not have some substantial mathematical connection. Sometimes puzzles or games have been invented that superficially concern math—usually arithmetic—but are often curiosities unsupported by any deeper theory. Mathematicians generally sneer at such mental exercises, likening the appeal of such problems to a shoot 'em up, car chase flick as opposed to an artistically crafted, Oscar-winning movie. There's no contention that part of the goal of recreational math is to have fun—that's recreational, right?—and even the best mathematicians might concede that what

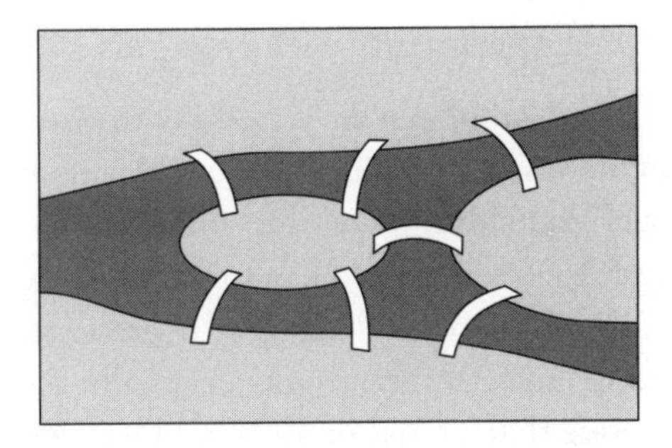

FIGURE 7.9: The Seven Bridges of Königsberg.

starts out as play can evolve into something rather sophisticated (and hopefully still fun).

Consider the old nursery rhyme, "As I was going to St Ives":

As I was going to St. Ives,
I met a man with seven wives,
Each wife had seven sacks,
Each sack had seven cats,
Each cat had seven kits:
Kits, cats, sacks, and wives,
How many were there going to St. Ives?

This conundrum could be seen as a precursor to the topic of geometric series. Mathematics has witnessed several nonserious or playful explorations that have swelled into robust research areas. Games of chance evolved into the theory of probability. Conway's Game of Life, explored in chapter 8, accelerated the development of cellular automata. Even Leibniz, the co-inventor of calculus, wrote in a 1715 letter, "Men are never more ingenious than in inventing games." Euler considered the Seven Bridges of Königsberg problem: Is it possible to take a walk so that each bridge is crossed exactly once (figure 7.9) ? This simple problem blossomed into the area of graph theory.

As a last ode to recreational mathematics, ponder a curious formula involving powers of seven:

$$1^7 + 4^7 + 4^7 + 5^7 + 9^7 + 9^7 + 2^7 + 9^7 = 14,459,929$$

Can you find a similar equation that uses a different power? HINT: Try the power 3. Solutions can be found in chapter 10.

Experiments with Integrals

Constructing a formal proof can be very different from the process in which the theorem was discovered. Mathematicians pride themselves on slick, aesthetic proofs, but sometimes the path to getting there is long and meandering, a process we'd like to forget. In published articles, the highly refined proof is typically all that is presented, depriving the reader of a sense of how the result was originally conjectured. Most mathematicians agree that a good proof should be economical, but one wonders whether clarity of craft should be sacrificed on the altar of economy. Even C. F. Gauss, the "Prince of Mathematics," deplored elaborating on his discovery process. Niels Abel said, "He [Gauss] is like the fox, who effaces his tracks in the sand with his tail" (Simmons, *Calculus Gems*, p. 177). Gauss defended his style, saying that "no self-respecting architect leaves the scaffolding in place after completing the building."

In practice, many mathematicians perform a huge amount of experimentation in their efforts to make conjectures. Their great ideas usually don't come out of nowhere; immersion into a problem often involves mucking around in a digital Petri dish. The burgeoning paradigm of experimental mathematics embraces computer algebra systems to explore new phenomena. Used effectively, the computer helps one to conjecture, test, refute, and even occasionally prove mathematical statements. Even Gauss admitted that his method of working was "through systematic experimentation" (*Journal of Experimental Mathematics*, Statement of Philosophy and Criteria). With the advent of the computer, numerical and symbolic tools have evolved to become a strong ally of the researcher.

Many fascinating results have been unearthed using experimental mathematics. Equation (2.1) (in chapter 2), the BBP formula for π, is a beautiful example where numerical and symbolic methods combined to produce a formula with new properties. Care, however, needs to be exercised; some apparent patterns turn out to be mirages. Consider the integral

$$\int_0^\infty \frac{\sin(x)}{x}dx = \frac{\pi}{2}$$

The function $\sin(x)/x$, sometimes called the *sinc* function, plays an important role in digital signal processing. Some similar integrals evaluate to the same value:

$$\int_0^\infty \frac{\sin(x)}{x}\frac{\sin(x/3)}{x/3}dx = \frac{\pi}{2}$$

$$\int_0^\infty \frac{\sin(x)}{x}\frac{\sin(x/3)}{x/3}\frac{\sin(x/5)}{x/5}dx = \frac{\pi}{2}$$

$$\int_0^\infty \frac{\sin(x)}{x}\frac{\sin(x/3)}{x/3}\frac{\sin(x/5)}{x/5}\frac{\sin(x/7)}{x/7}dx = \frac{\pi}{2}$$

$$\int_0^\infty \frac{\sin(x)}{x}\frac{\sin(x/3)}{x/3}\frac{\sin(x/5)}{x/5}\frac{\sin(x/7)}{x/7}\frac{\sin(x/9)}{x/9}dx = \frac{\pi}{2}$$

$$\int_0^\infty \frac{\sin(x)}{x}\frac{\sin(x/3)}{x/3}\frac{\sin(x/5)}{x/5}\frac{\sin(x/7)}{x/7}\frac{\sin(x/9)}{x/9}\frac{\sin(x/11)}{x/11}dx = \frac{\pi}{2}$$

$$\int_0^\infty \frac{\sin(x)}{x}\frac{\sin(x/3)}{x/3}\frac{\sin(x/5)}{x/5}\frac{\sin(x/7)}{x/7}\frac{\sin(x/9)}{x/9}\frac{\sin(x/11)}{x/11}\frac{\sin(x/13)}{x/13}dx = \frac{\pi}{2}$$

It's tempting to speculate that these seven integrals establish a general pattern. However, the computer returns a different value for the next integral:

$$\int_0^\infty \frac{\sin(x)}{x}\frac{\sin(x/3)}{x/3}\frac{\sin(x/5)}{x/5}\frac{\sin(x/7)}{x/7}\frac{\sin(x/9)}{x/9} \times \frac{\sin(x/11)}{x/11}\frac{\sin(x/13)}{x/13}\frac{\sin(x/15)}{x/15}dx$$

$$= \frac{467,807,924,713,440,738,696,537,864,469}{935,615,849,440,640,907,310,521,750,000}\pi$$

$$\approx 0.499999999992647\pi$$

A researcher who found this result suspected that there was a bug in the software. The truth, however, is that the integral is correct. The explanation for this change is a bit technical, but the critical reason is that $\frac{1}{3} + \frac{1}{5} + \cdots + \frac{1}{13} < 1$, whereas adding the next term $\frac{1}{15}$ pushes the sum over 1, making a difference in the value of the integral. This example gives a cautionary tale to those who rush too quickly to make a sweeping conclusion. An apocryphal statement attributed to the famous economist John Maynard Keynes is, "When the facts change, I change my mind. What do you do, sir?"

8

The Number Eight

Telephone operators should work eight hours and sleep eight hours—but not the same eight hours.
—28 September 1921, Fort Wayne (IN) News-Sentinel

The number 8 weaves its web in various places, from the tasty Pizza Theorem to the playful Game of Life and the towering heights of E_8. Like an octopus's hold, the number 8 will grip you with perfect card shuffling and the beautiful Sierpiński Carpet. Enjoy!

The Pizza Theorem

Jacob and Lucas have baked a circular pizza and are ready to devour it. Jacob asks Lucas to cut the pizza into eight equal slices by first marking the center then making the four slices through their center, each 45° apart. Lucas, being a ravenous teenager, has made the correct cuts, but was substantially off when he identified the center, resulting in both troll- and hobbit-sized slices. Lucas thought he was behind the eight ball, but soon a smile played on his face. The truth of the Pizza Theorem dawned upon him: if each of them takes alternating slices, they would each have the same amount of pizza. A beautiful proof by picture is in figure 8.1.

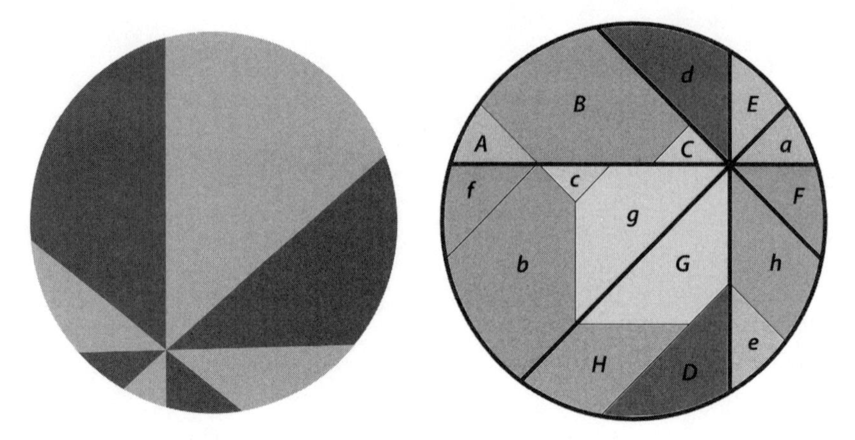

FIGURE 8.1: The Pizza Theorem and a proof.

Shuffling Cards

How many times should one mix a standard deck of cards to ensure that they are well shuffled? For most people, three or four shuffles seems adequate; few people would shuffle more than this. Most serious card players, however, know that this is insufficient. Capitalizing on this situation, clever gamblers and bridge players leverage their knowledge of the preshuffled deck to glean information about the partially mixed deck. And they win more.

So how many shuffles are needed to really mix the cards? We should mention that by "mixing" we mean a rough riffle shuffle. In 1992, researchers studied this problem with computer simulations and conjectured that seven shuffles is enough to mix the cards well. Subsequently, they produced a careful mathematical proof and also argued that any further shuffling does not significantly improve the mixing.

What if, however, we do *perfect* shuffles? A perfect riffle shuffle—sometimes called a *Faro shuffle*—cuts the deck exactly in half and the shuffle perfectly alternates the cards with the top card staying on top. For example, a perfect shuffle permutes the cards numbered $\{1, 2, 3, 4, 5, 6, 7, 8\}$ to $\{1, 5, 2, 6, 3, 7, 4, 8\}$. It can be shown that eight consecutive perfect shuffles leaves a standard 52-card deck just like it started. To see this mathematically, note that if a card starts in position k, a Faro shuffle takes it to position $2k - 1$ if $k \leq 26$ and $2k - 52$ if

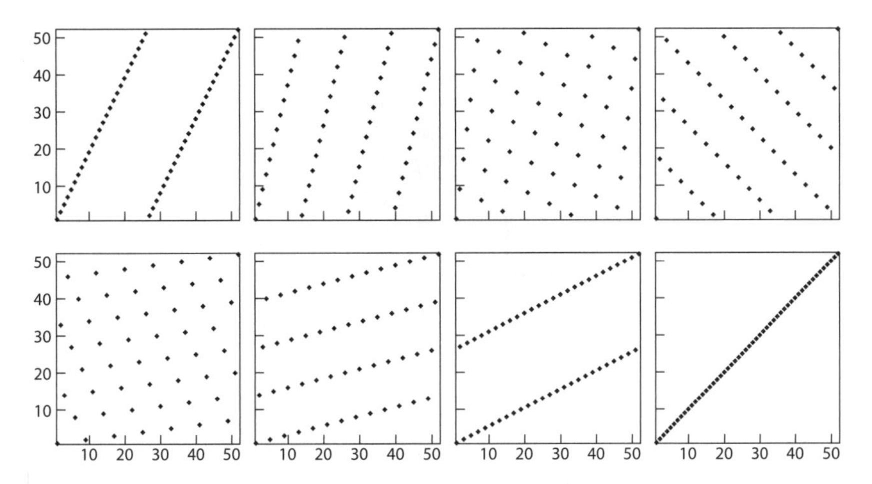

FIGURE 8.2: Eight consecutive shuffles of a deck of cards. The horizontal axis is the card number and the vertical axis is its position.

$k > 26$. For example, the card starting in the fifth position wanders as follows:

$$5 \rightarrow 9 \rightarrow 17 \rightarrow 33 \rightarrow 14 \rightarrow 27 \rightarrow 2 \rightarrow 3 \rightarrow 5$$

To see how the whole deck migrates, see figure 8.2.

The Game of Life

The world of classical applied mathematics is often written in terms of differential equations, an outgrowth of calculus. The use of such equations has been spectacularly successful at modeling various phenomena, such as fluid flow, elasticity, and cosmology. At the core of this approach, one assumes that certain governing laws of physics work at an infinitesimal—think submicroscopic—level.

In many phenomena, however, the governing laws only make sense at a macroscopic level. While differential equations have been used to study traffic flow and population growth, the basic objects in these scenarios—cars and people—are discrete objects. It may make more sense to develop and understand laws that work in a discrete scenario. Does a car slow down based only on the car immediately in front of it?

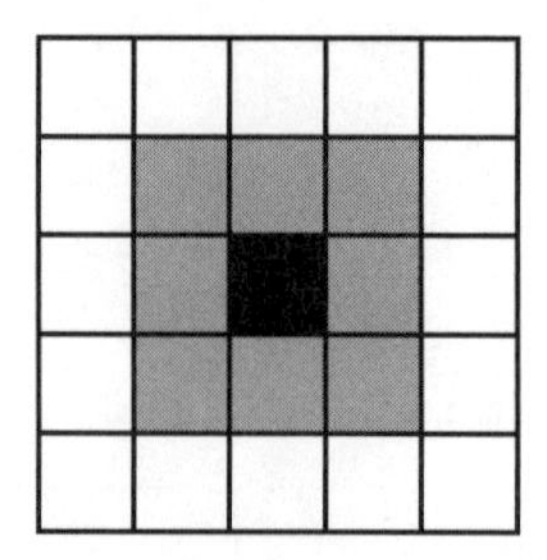

FIGURE 8.3: The Moore neighborhood for the Game of Life.

What factors affect human reproduction rates? In mathematical models of swarms—this could be fish, birds, insects, etc.—mathematicians have used local (think "nearby") properties in their attempts to model the swarm's movement. This includes moving in the same direction as your neighbors, remaining close to them, but also avoiding collisions. In the 1940s, mathematical giants Stanislaw Ulam and John von Neumann studied these discrete, neighborhood models in the abstract. Such models are called *cellular automata* and have also been studied by computer scientists and theoretical biologists. A cellular automaton (CA) consists of a grid made up of cells, each of which can be in a finite number of states. Once the rules for the evolution of a CA are specified, one can study the long-term behavior of such a system.

With the idea of reproduction in mind, von Neumann was interested in a CA where parts of an initial state could reproduce themselves. Though he created a complicated example that did what he wanted, broader attention to CA ensued only when John Conway developed what is now called the Game of Life (not to be confused with the board game developed by Milton Bradley). Devised in 1970, Conway's Game of Life became widely known when it appeared in the October 1970 issue of *Scientific American* in Martin Gardner's "Mathematical Games" column.

The Game of Life has very simple rules yet can display fantastically complex phenomena. On a rectangular grid, each cell has eight neighbors; this is sometimes called the *Moore neighborhood* of a cell (figure 8.3).

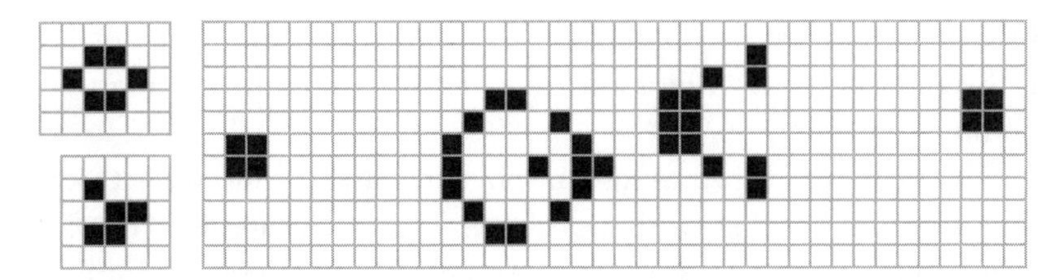

FIGURE 8.4: Game of Life configurations: the beehive, glider, and four parts of a glider gun.

In each generation, a cell is either on or off—alive or dead. The status of each cell in the next generation is determined by its current state and that of its neighbors:

- If a live cell has fewer than two live neighbors, it dies (underpopulation).
- If a live cell has two or three live neighbors, it lives.
- If a live cell has greater than three live neighbors, it dies (overpopulation).
- If a dead cell has exactly three live neighbors, it comes to life (reproduction).

Fascination with various patterns generated by the Game of Life caught on immediately. This includes patterns (figure 8.4) that do not change (the "block," the "boat," and the "beehive"), repeat after a finite number of iterations (the "blinker" and the "toad"), or replicate themselves in a shifted position and thus seem to move across the grid (the "glider" or the "spaceship"). Some patterns can mushroom into huge monsters but eventually collapse into nothing. Conway conjectured that no finite pattern could grow indefinitely. This suspected restriction was emancipated by the construction of a "glider gun," a pattern that oscillates itself and regularly pumps out gliders, which march off to infinity. Another example was a "puffer train," which, like the glider, replicated itself in a shifted position but also left behind it debris.

For theoretical computer scientists, the Game of Life offers something stronger than just nice patterns. By using gliders in creative ways to create and destroy other objects, logic gates such as AND,

OR, and NOT can be simulated. This means that the Game of Life is theoretically as versatile as any computer with unlimited memory and no time constraints.

The Game of Life has also inspired researchers in chaos theory, philosophy, and biology with regard to emergence, an area where simple rules can lead to complex patterns. An example of emergence from the natural world is ants. Each individual ant carries out duties by picking up stimuli from nearby ants and other local situations, not by some heirarchical commands given by the queen or the colony. This begs the question: Are there simple rules in the natural world that lead to self-organizing systems?

Repetition in Pascal's Triangle

After centuries of analysis, Pascal's triangle continues to reveal patterns and offer conundrums. One of these problems concerns repeated values. Of course, the number 1 appears infinitely often in Pascal's triangle; it forms the border on both sides. Any other number n appears at most a finite number of times since it cannot appear beyond the nth row.

Any number in Pascal's triangle that is *not* in the outer two layers will appear at least three times, usually four. For example,

$$15 = \binom{6}{2} = \binom{6}{4} = \binom{15}{1} = \binom{15}{14}$$

It's trickier to find numbers that appear more often. Examples of six appearances include

$$120 = \binom{120}{1} = \binom{120}{119} = \binom{16}{2} = \binom{16}{14} = \binom{10}{3} = \binom{10}{7}$$

and

$$1{,}540 = \binom{1{,}540}{1} = \binom{1{,}540}{1{,}539} = \binom{56}{2} = \binom{56}{54} = \binom{22}{3} = \binom{22}{19}$$

David Singmaster—famous for having developed a widely used algorithm for solving Rubik's Cube—showed that there are infinitely many numbers that appear at least six times in Pascal's triangle. For some positive integer i, let $N = \binom{F_{2i+2}F_{2i+3}}{F_{2i}F_{2i+3}}$, where F_n is the nth Fibonacci number. Then

$$\binom{N}{1} = \binom{N}{N-1} = \binom{F_{2i+2}F_{2i+3}}{F_{2i}F_{2i+3}} = \binom{F_{2i+2}F_{2i+3}-1}{F_{2i}F_{2i+3}+1}$$
$$= \binom{F_{2i+2}F_{2i+3}}{F_{2i+1}F_{2i+3}} = \binom{F_{2i+2}F_{2i+3}-1}{F_{2i+1}F_{2i+3}-2}$$

The number 3,003 appears eight times:

$$3,003 = \binom{3,003}{1} = \binom{3,003}{3,002} = \binom{78}{2} = \binom{78}{76}$$
$$= \binom{15}{5} = \binom{15}{10} = \binom{14}{6} = \binom{14}{8}$$

Singmaster conjectured that there is an upper bound on the multiplicity of any number $n > 1$ in Pascal's triangle. He thought that it might be 10 or 12, but no example with multiplicity larger than 8 has been found.

The Sierpiński Carpet

The 1980s was a real coming-out time for the area of fractals. The realization that fractals could be generated with very simple computer code, coupled with the rise of the personal computer, led to fractals being embraced by professionals and amateurs alike. One of these popular images was the Sierpiński Carpet. Starting with a square, divide it into nine equal subsquares. Now remove the middle subsquare. For each of the remaining eight subsquares, perform the same operations, over and over. The limiting geometrical image is called the *Sierpiński Carpet* (figure 8.5). The object has zero area but infinite boundary. The self-similarity in fractals can be beneficial for antenna design.

The Sierpiński Carpet can be considered a two-dimensional generalization of the Cantor set encountered in chapter 1. By writing points in the square $[0,1) \times [0,1)$ in base 3, the Sierpiński Carpet excludes the points that have the digit 1 in the same position. The Sierpiński Carpet has often been overshadowed by its triangular sibling,

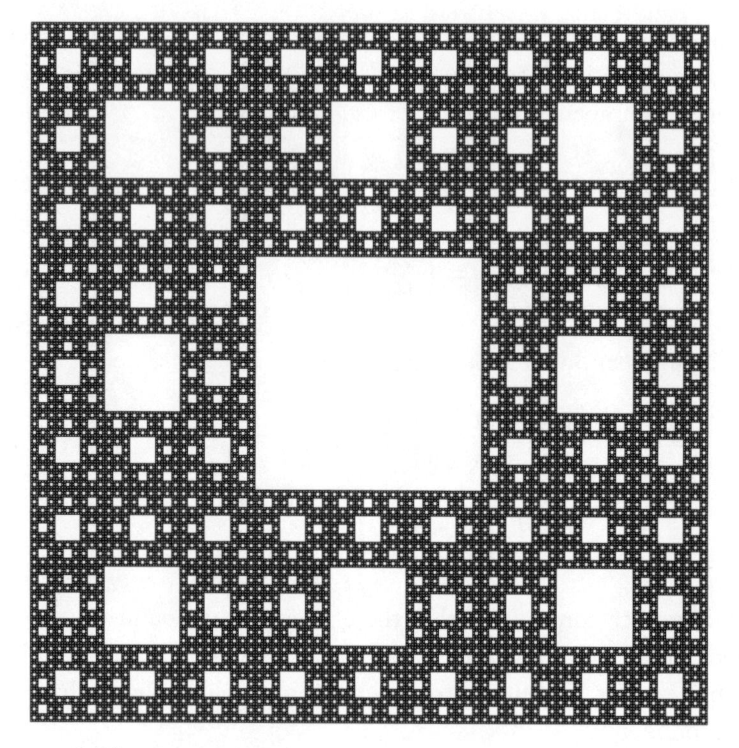

FIGURE 8.5: The Sierpiński Carpet.

the Sierpiński Gasket. However, the carpet has a special topological property: Any one-dimensional graph can be found as an image within the carpet. For example, any stick figure or tree diagram can be found, with some bending and scaling, in the Sierpiński Carpet. In this sense, the Sierpiński Carpet is *universal.*

Quaternions and Octonions

From the quadratic formula for solving the equation $ax^2 + bx + c = 0$, one naturally sees how the complex numbers could arise. Although the square root of a negative number may seem jarring at first glance, there is a beauty to expanding the real numbers to this broader system. We can manipulate complex numbers with simple operations such as addition, subtraction, multiplication, division, square roots, exponentials, logarithms, etc., and the result is always a complex

Table 8.1
Quaternion Multiplication Table

×	1	i	j	k
1	1	i	j	k
i	i	-1	k	$-j$
j	j	$-k$	-1	i
k	k	j	$-i$	-1

number. The basic formula for multiplication,

$$(a + ib) \cdot (c + id) = (ac - bd) + i(bc + ad)$$

can be multiplied by its conjugate to produce the the Brahmagupta–Fibonacci identity seen in chapter 2:

$$(a^2 + b^2)(c^2 + d^2) = (ac - bd)^2 + (ad + bc)^2 \tag{8.1}$$

for any real numbers a, b, c, and d.

If we add another "number" besides 1 and i, could we produce an even larger space with interesting properties? The Irish mathematician William Rowan Hamilton sought such a space for many years but failed to find one. Out of the ashes, however, rose his idea for a *four-dimensional* algebra, which he promptly called the *quaternions*. In this algebra, there are three distinct imaginary numbers i, j, and k such that $i^2 = j^2 = k^2 = -1$. All the possible products of the four elements $\{1, i, j, k\}$ are listed in table 8.1.

Any element x that is a quaternion can be represented as $x = a + bi + cj + dk$, where a, b, c, and d are real numbers. Table 8.1 is used to multiply two quaternions:

$$\begin{aligned} x_1 \cdot x_2 &= (a_1 + b_1 i + c_1 j + d_1 k) \cdot (a_2 + b_2 i + c_2 j + d_2 k) \\ &= a_1 \cdot (a_2 + b_2 i + c_2 j + d_2 k) + b_1 i \cdot (a_2 + b_2 i + c_2 j + d_2 k) \\ &\quad + c_1 j \cdot (a_2 + b_2 i + c_2 j + d_2 k) + d_1 k \cdot (a_2 + b_2 i + c_2 j + d_2 k) \\ &= (a_1 a_2 - b_1 b_2 - c_1 c_2 - d_1 d_2) + (a_1 b_2 + b_1 a_2 + c_1 d_2 - d_1 c_2) i \\ &\quad + (a_1 c_2 - b_1 d_2 + c_1 a_2 + d_1 b_2) j + (a_1 d_2 + b_1 c_2 - c_1 b_2 + d_1 a_2) k \end{aligned} \tag{8.2}$$

If you haven't noticed already, the order in which quaternions are multiplied makes a difference. For example, table 8.1 indicates that $i \cdot j = -j \cdot i$. When the order of multiplication makes a difference, we say that the algebra is *noncommutative*.

The last three components of a quaternion form the the so-called vector part. Because this part is three-dimensional, the structure of quaternions can be used to describe the geometry of three-dimensional space. For example, the convoluted structure of the cross-product—a product used to multiply two three-dimensional vectors together—is directly related to the quaternionic structure.

By multiplying equation (8.2) by its conjugate, one arrives at Euler's four-square identity witnessed in chapter 4:

$$
\begin{aligned}
&(a_1^2 + b_1^2 + c_1^2 + d_1^2) \cdot (a_2^2 + b_2^2 + c_2^2 + d_2^2) \\
&\quad = (a_1a_2 - b_1b_2 - c_1c_2 - d_1d_2)^2 + (a_1b_2 + b_1a_2 + c_1d_2 - d_1c_2)^2 \\
&\quad +(a_1c_2 - b_1d_2 + c_1a_2 + d_1b_2)^2 + (a_1d_2 + b_1c_2 - c_1b_2 + d_1a_2)^2
\end{aligned}
$$

Soon after Hamilton published his work on quaternions, his friend John T. Graves discovered what we now call the *octonions*, an 8-dimensional algebra. Every octonion x can be written as

$$x = x_0e_0 + x_1e_1 + x_2e_2 + x_3e_3 + x_4e_4 + x_5e_5 + x_6e_6 + x_7e_7$$

where the coefficients $\{x_k\}$, $0 \le k \le 7$, are real and the unit elements $\{e_k\}$, $0 \le k \le 7$, are distinct. The multiplication table for the octonions is in table 8.2. An aid to remembering how to multiply the units $e_1, e_2, \cdots, e_7$ is to use the Fano plane encountered in chapter 7; see figure 8.6. If a is connected to b by an arrow, then $a \cdot b = c$ where c is the unique unit such that a, b, and c are collinear.

Like the quaternions, one sees that the octonions are noncommutative. They are stranger still; the octonions are nonassociative. This means that there are examples where $(x_1 \cdot x_2) \cdot x_3 \neq x_1 \cdot (x_2 \cdot x_3)$. For example, $(e_1 \cdot e_4) \cdot e_3 = e_5 \cdot e_3 = e_6$, whereas $e_1 \cdot (e_4 \cdot e_3) = e_1 \cdot (-e_7) = -e_6$. Despite having weaker structural properties, the same process applied to the previous algrebras carries over here: expand the

Table 8.2
Octonion Multiplication Table

$\times$	e_0	e_1	e_2	e_3	e_4	e_5	e_6	e_7
e_0	e_0	e_1	e_2	e_3	e_4	e_5	e_6	e_7
e_1	e_1	$-e_0$	e_3	$-e_2$	e_5	$-e_4$	$-e_7$	e_6
e_2	e_2	$-e_3$	$-e_0$	e_1	e_6	e_7	$-e_4$	$-e_5$
e_3	e_3	e_2	$-e_1$	$-e_0$	e_7	$-e_6$	e_5	$-e_4$
e_4	e_4	$-e_5$	$-e_6$	$-e_7$	$-e_0$	e_1	e_2	e_3
e_5	e_5	e_4	$-e_7$	e_6	$-e_1$	$-e_0$	$-e_3$	e_2
e_6	e_6	e_7	e_4	$-e_5$	$-e_2$	e_3	$-e_0$	$-e_1$
e_7	e_7	$-e_6$	e_5	e_4	$-e_3$	$-e_2$	e_1	$-e_0$

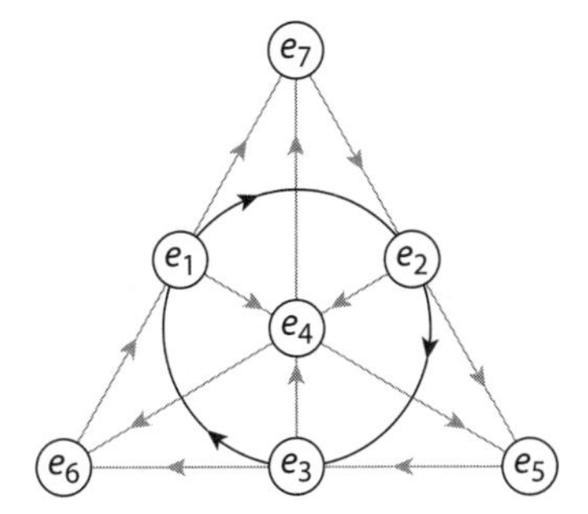

FIGURE 8.6: Using the Fano plane to multiply octonions.

product of two arbitrary octonions, then multiply this equation by its conjugate. This produces a formula where the product of two sums of eight squares equals the sum of eight squares. This equation was discovered independently around 1818 by Ferdinand Degen and is known as Degen's eight-square identity.

Adolf Hurwitz proved that only four spaces exhibit the multiplicative structure we've discussed: the reals, complex numbers, quaternions, and octonions. Starting with the real numbers, one sees that every expansion to a larger space loses some structure: the complex numbers are not ordered, the quaternions are not commutative, and the octonions are not associative. John Baez captured this breakdown in structure by comparing these algebras to characters in a family:

> The real numbers are the dependable breadwinner of the family, the complete ordered field we all rely on. The complex numbers are a slightly flashier but still respectable younger brother:

> not ordered, but algebraically complete. The quaternions, being noncommutative, are the eccentric cousin who is shunned at important family gatherings. But the octonions are the crazy old uncle nobody lets out of the attic: they are nonassociative (Baez, "The Octonians," p. 145).

The octonions languished for many years because they apparently lacked connection to mathematical physics. Glimpses appeared in the 1930s, but only in the 1980s was the tie made to the area of string theory.

The Summit of E_8

We have seen that mathematicians study the symmetries of an object by identifying its underlying structure. For example, a cube can be rotated about its vertical axis a quarter turn and its eight vertices occupy the same set of positions. The quarter turn could be applied again to produce a new transformation. Rotations around other axes also exist, and these can be combined. The set of transformations that leave the cube's vertices unchanged is an example of a *group*. The study of groups has boomeranged back to shed light on many areas of pure mathematics, crystallography, and mathematical physics.

The group representing the symmetries of the cube is a finite group since there are only finitely many ways to transform the vertices. Of a different nature is the symmetry group of a wine bottle. The bottle can be rotated about its axis by any angle. The corresponding group of rotations is infinite in size and is represented by the circle. This is a simple example of a *Lie group* (pronounced "lee" after Norwegian mathematician Sophus Lie). Lie groups provide a natural framework for many parts of theoretical physics. Examples include the Heisenberg group used in quantum mechanics and the gauge group, a 12-dimensional group that represents the so-called standard model in particle physics since it uses 1 photon, 3 vector bosons, and 8 gluons.

Of special interest are compact Lie groups, groups whose symmetries are, in a technical sense, bounded. The Lie group representing the symmetries of the bottle is such an example. In fact, most compact Lie groups belong to one of four infinite collections. These are often used to explain, for example, the symmetries witnessed in spherical and projective geometry. Five Lie groups, however, do not belong to any one of these collections. Labeled as E_6, E_7, E_8, F_4, and G_2, they are called the *exceptional Lie groups*. The group E_8 is the largest, with 248 dimensions, and has recently been the focus of much attention.

The group E_8 can be constructed using the so-called E_8 lattice. This is a web of points made from 8-dimensional vectors (called *roots*) whose entries have the following properties:

- the entries are either all integers or all integers plus $1/2$,
- the sum of the entries is an even number, and
- the sum of the squares of the entries equals 2.

Examples of roots include the vectors $(1, 0, 0, -1, 0, 0, 0, 0)$ and $(1/2, 1/2, -1/2, 1/2, 1/2, 1/2, 1/2, -1/2)$. There are 240 different roots. The lattice formed from these roots is sometimes called a "diamond lattice," and each "cell" has volume equal to 1 (such a lattice is called *unimodular*). The dimension of the group is simply the sum $8 + 240 = 248$ because there are 240 roots and there are eight degrees of freedom for each root.

A project known as the Atlas of Lie Groups and Representations was initiated in 2007. One of its goals was to understand the representations of the Lie group E_8, which are based on the E_8 lattice (figure 8.7). A representation is a way to understand some symmetry of a group using matrices. For example, the circle encountered with the wine bottle has rotation matrices as representations. To understand the representations in complicated Lie groups, it is enough to restrict attention to the irreducible representations, basic representations that play the same role that the primes play among the positive integers. After four years of work by about 20 mathematicians and 77 hours of supercomputer time, they cracked E_8. The team found that there are 453,060 irreducible representations, and they mapped the connections between each pair of representations in a $453,060 \times 453,060$ matrix. For comparison, it has been noted that the amount of information generated dwarfs the results of the Human Genome Project, which mapped the genetic information of the human genome.

Why all this fuss over one group? E_8 is connected to string theory, a modern paradigm of the physical reality in which we live. Heterotic

FIGURE 8.7: A projection of E_8.

string theory claims that we live in 26-dimensional space, not the mundane 4-dimensional space (three "space" dimensions plus time). In order to reduce the number of dimensions from 26 to the perceived 4, part of the theory claims that there must be a 16-dimensional space that curls up on itself in a nice way. It ends up that there are only two ways this can happen mathematically, and explaining one of them involves E_8.

9

The Number Nine

> Later outside, in the appropriately numerically named Times Square, I gazed up at the towering skyscrapers and felt surrounded by 9s — the number I most associate with feelings of immensity.
>
> — Daniel Tammet, *Born on a Blue Day*

> If I read in an article that a person felt intimidated by something, I imagine myself standing next to the number 9.
>
> — Daniel Tammet, *Born on a Blue Day*

The last number in this book, 9, connects back to several topics we've already encountered, including prime numbers, packings, and powers of numbers. The Heegner numbers may push you farther with their surprising connections, but you don't have to be a cat with nine lives to survive and thrive on this stunning mathematics.

Nine Points and Collinearity

There are two theorems that, starting with six points, produce three new points and force them to be collinear. The older of the two results is Pappus's Hexagon Theorem. Suppose that there are two lines, each containing three points, and six connecting lines are made that result

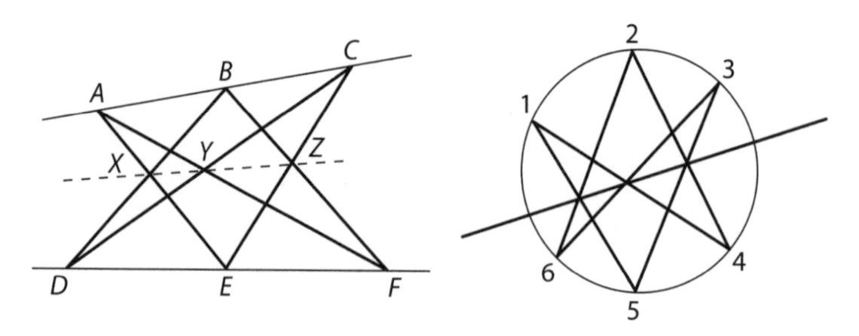

FIGURE 9.1: Pappus's Hexagon Theorem (left) and Pascal's Theorem (right).

in three "middle" points of intersection. The theorem asserts that the three new points must be collinear (figure 9.1).

The other result is Pascal's Theorem, also known by a cosmic title as the Hexagrammum Mysticum Theorem. Suppose that one has a hexagon inscribed in a conic section. For each of the three pairs of opposite sides, form the lines through the sides to produce three intersection points. The theorem claims that these three points are also collinear (figure 9.1). Pascal's Theorem generalizes Pappus's Hexagon Theorem because one can take ellipses of increasingly larger size while keeping the points in roughly the same locations. As the ellipse becomes arbitrarily large, each set of three points on the ellipse becomes collinear.

Pascal's Theorem is itself a special case of a more general result that goes the whole nine yards. This theorem involves cubic curves, that is, curves defined by polynomial equations whose degree equals three. Such curves were intensely studied during the seventeenth century near the genesis of calculus. Some of these curves have colorful names such as the conchoid of de Sluze, the folium of Descartes, and the witch of Agnesi. The more general result is the Cayley–Bacharach Theorem. Let A and B be two cubic curves that intersect at nine points. If C is a cubic curve that passes through eight of the points, then it passes through the ninth point as well.

These theorems illustrate another force that propels mathematicians: generalization. The idea is that sometimes some mathematical fact is simply a special case of a more general principle. Usually the general result works because of some "deeper" mathematics. The search for truth marches on.

Casting Out Nines

The practice of detecting errors in a text or message has a history going back thousands of years. An early success involved copying portions of the Bible. Jewish scribes computed quantities such as the number of words per line, characters per line, and words per page. Comparing the counts of a copy to the original was a relatively efficient way to offer modest assurance that the copying was accurate. Just one mistake was enough to throw out a page of a scribe's work. The accuracy of the Dead Sea Scrolls, discovered in the mid-twentieth century, confirmed the effectiveness of these careful checks.

Besides checking text, assessing the accuracy of numerical calculations has also received attention. One can imagine that this was particularly relevant when a "computer" referred to a person, not a machine. What is an easy way to measure whether a calculation is consistent? Such a check is sometimes called a *sanity test*. Enter the test "casting out nines" (CON).

The CON method goes back at least one millennium. Let's start with a simple example. Suppose the claim is that $1,382 \times 2,596 = 3,587,672$. For each number involved, we want to add the digits. If the sum is greater than nine, add the digits again, and keep doing this until only one digit is left. So adding the digits of $1,382$, we get 14, and adding those digits leaves 5. Similarly, $2,596$ transforms to 22 and then to 4. Lastly, $3,587,672$ becomes 38, then 11, and finally 2. Now comes the check: the product of 5 and 4—the leftovers of $1,382$ and $2,596$, respectively—is 20, which transforms to 2, the same as the transformed version of $3,587,672$.

It's important to say that this check does not prove that the original product is true; it simply claims that it is possible that the product is correct. If we had claimed that the product incorrectly equaled $3,542,672$, this number also transforms to 2. The test is best at catching when a mistake is made, though it can still miss. Note also that the CON method involves only a few simple additions, computationally much cheaper than checking the whole product again.

So how does this method work? And where are the nines? If $ab = c$, then for any number r, we have

$$((a \bmod r) \times (b \bmod r)) \bmod r = (c \bmod r)$$

When $r = 9$, these expressions collapse beautifully. For example,

$$\begin{aligned} 1{,}382 &= (1 \times 1{,}000) + (3 \times 100) + (8 \times 10) + 2 \\ &= [1 \times (1 + 999)] + [3 \times (1 + 99)] + [8 \times (1 + 9)] + 2 \\ &= (1 + 3 + 8 + 2) + 9M \end{aligned}$$

where M is some integer whose exact value is irrelevant. By "casting out" the nines, we are left with $1 + 3 + 8 + 2 = 14$. Casting out the nines from 14 leaves 5. Since any power of 10 reduces to 1, this shows how simply adding the digits does the job.

Primes and Nines

We've already seen that there are many open challenges concerning prime numbers. In attempts to resolve these problems, many partial results using modern methods connect these conjectures with the number nine.

ODD PERFECT NUMBERS

We saw in chapter 2 that there are no known odd perfect numbers. Recall that these are odd numbers whose sum of divisors equals twice the number. If an odd perfect number were to exist, then it must exceed 10^{1500} and have at least 75 prime factors. More interesting for the moment, it has been shown that at least nine of these prime factors must be distinct.

TWIN PRIMES

Notwithstanding the very recent progress toward a proof of the Twin Prime Conjecture, it is unclear whether this approach will yield a complete solution. Brun's Theorem—the result that the sum of the reciprocals of all the twin primes is finite—is one of the fascinating results in this area of study. Brun had a lesser known result concerning this conundrum: there exist infinitely many n such that n and $n + 2$ each has at most nine prime factors.

GETTING PRIMES CLOSE TOGETHER

In chapter 2, we encountered Bertrand's Postulate: for any integer $n \geq 2$, there is at least one prime between n and $2n$. Some experimentation suggests that for large values of n, many primes appear in this gap. A much stronger result states that if $x > \pi$, then there are at least nine primes between x^3 and $(x+1)^3$. Note that if x is large, then $(x+1)^3/x^3 \approx 1$, as opposed to the ratio $(2n)/n = 2$ in Bertrand's Postulate. There's only one caveat to this more potent result; its proof requires that the Riemann Hypothesis is true, the most well known of the Millennium Prize Problems.

The Fifteen Theorem

As we saw in chapter 4, Lagrange's Four-Square Theorem asserts that every positive integer n can be written as the sum of four squares, that is, $n = a^2 + b^2 + c^2 + d^2$ for some integers a, b, c, and d. The expression on the right is an example of a *quadratic form* and can also be written as $v^T M v$ where v^T is the row vector (a, b, c, d) and M is the identity matrix

$$M = \begin{bmatrix} 1 & 0 & 0 & 0 \\ 0 & 1 & 0 & 0 \\ 0 & 0 & 1 & 0 \\ 0 & 0 & 0 & 1 \end{bmatrix}$$

This quadratic form is *positive definite* because it equals zero if and only if $a = b = c = d = 0$, and is positive otherwise.

In an attempt to generalize Lagrange's result, John H. Conway and W. A. Schneeberger proved in 1993 the *Fifteen Theorem*: if a positive definite quadratic form has a corresponding matrix whose entries are all integers and the form itself attains all the values from 1 to 15, then the form can attain all positive integers (we say this form is *universal*). A tighter version of the theorem claims that if the nine values 1, 2, 3, 5, 6, 7, 10, 14, and 15 are attained, then the form is universal. This theorem is sharp—none of the nine values from this list can be removed—since one can produce a quadratic form that hits every positive integer except a specified number from the list. For example, the form

$$a^2 + 2b^2 + 5c^2 + 5d^2$$

whose matrix representation is

$$M = \begin{bmatrix} 1 & 0 & 0 & 0 \\ 0 & 2 & 0 & 0 \\ 0 & 0 & 5 & 0 \\ 0 & 0 & 0 & 5 \end{bmatrix}$$

represents all values except 15.

Conway and Schneeberger's proof was complicated and never formally published. Manjul Bhargava found a simpler proof in 2000 and listed all the possible universal forms (204 of them). Indeed, Bhargava and Jonathan P. Hanke found an interesting related result. Instead of requiring that the matrix have integer entries, suppose we require the weaker condition that the form is integer-valued, that is, the coefficients in the form are integers. One sees that this is a less demanding condition since $x^2 + xy + y^2$ is a positive definite quadratic form but its corresponding matrix is

$$M = \begin{bmatrix} 1 & 1/2 \\ 1/2 & 1 \end{bmatrix}$$

The new result claims that the integer-valued quadratic form is universal if it hits the 29 values 1, 2, 3, 5, 6, 7, 10, 13, 14, 15, 17, 19, 21, 22, 23, 26, 29, 30, 31, 34, 35, 37, 42, 58, 93, 110, 145, 203, and 290. This result is sometimes called the 290 Theorem.

Circle Packings with Two Sizes

A manufacturer of tomato-based products is interested in lowering shipping costs. Most of the buyers of canned tomatoes also order cans of tomato paste, whose radius is smaller. If each product is packaged in cases using a hexagonal lattice as seen in chapter 6, the density of

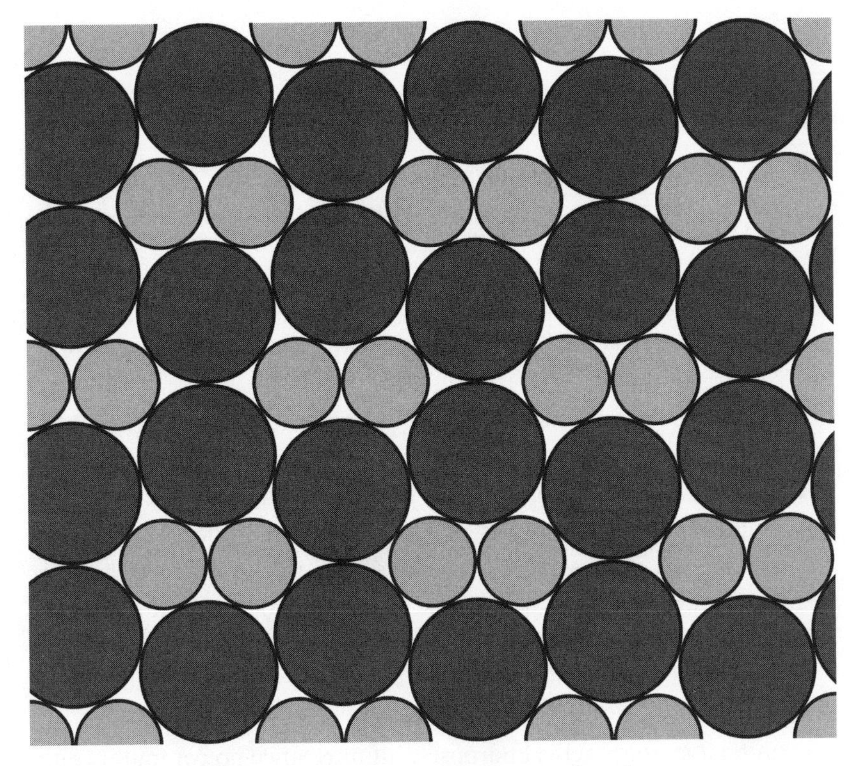

FIGURE 9.2: A tight packing with two different-sized circles.

the cans—the proportion of used space to the whole space—is roughly 0.9069. However, it's possible to pack two different-sized cans with a higher density. It ends up that there are exactly nine possible circle packings with two different-sized circles that are tight (no circle can wiggle around). The simplest packing is shown in figure 9.2.

Catalan's Conjecture

Among the constellation of natural numbers, perfect powers have singled themselves out as being useful in all areas of mathematics. From the first few—1, 4, 8, 9, 16, 25, 27, 32, and 36—it is not clear how large or small the successive gaps may be. If a_n denotes the nth perfect power, a nice formula proved by Christian Goldbach confirms that these numbers spread out:

$$\sum_{n=2}^{\infty} \frac{1}{a_n - 1} = 1$$

More recently, it was shown that

$$\sum_{n=2}^{\infty} \frac{1}{a_n + 1} = \frac{\pi^2}{3} - \frac{5}{2}$$

In 1844, Eugène Charles Catalan conjectured that 9 is the only perfect power that differs from another perfect power by exactly one, that is, $3^2 - 2^3 = 1$. The special case when the bases are 2 and 3 was proved by Gersonides in 1343. Using advanced tools from number theory, it was proven in the 1970s that if there is another solution, then the bases cannot exceed the whopping number $B = \exp(\exp(\exp(\exp(730))))$, a number whose size defies comprehension (just the number exp(730) absolutely dwarfs the estimated number of atoms in the universe, 10^{80}; the size of B is truly staggering). Although it is finite, it is, for practical purposes, infinite, since no computer could ever check the numbers up to this point. In 2002, the mathematical world was caught off guard when Preda Mihăilescu proved Catalan's Conjecture.

An interesting sequel, which combines both Fermat's Last Theorem and the Catalan Conjecture, is the unimaginatively named Fermat–Catalan Conjecture. It claims that there are only a finite number of integer solutions to the equation

$$x^p + y^q = z^r \tag{9.1}$$

where the bases x, y, and z are greater than 1, and the exponents satisfy

$$\frac{1}{p} + \frac{1}{q} + \frac{1}{r} < 1 \tag{9.2}$$

Faltings' Theorem implies that for any single choice of exponents (p, q, r) satisfying inequality (9.2), there are only finitely many solutions to equation (9.1). The Fermat–Catalan Conjecture, which seems exceedingly difficult to solve, asserts that over *all* possible choices of the exponents, the total number of solutions is still finite. Currently, there are only nine known solutions:

$$2^5 + 7^2 = 3^4$$

$$7^3 + 13^2 = 2^9$$

$$2^7 + 17^3 = 71^2$$

$$3^5 + 11^4 = 122^2$$

$$17^7 + 76,271^3 = 21,063,928^2$$

$$1,414^3 + 2,213,459^2 = 65^7$$

$$9,262^3 + 15,312,283^2 = 113^7$$

$$43^8 + 96,222^3 = 30,042,907^2$$

$$33^8 + 1,549,034^2 = 15,613^3$$

Note that for each of the nine solutions, one of the exponents is a 2. This steers us to Beal's Conjecture: if there is a solution to equation (9.1) and $p, q, r > 2$, then x, y, z must have a common factor. Andrew Beal, a wealthy Dallas banker with an interest in number theory, offered in 1997 a prize of $5,000 for a solution to his problem. He stipulated that the prize would grow each year by $5,000, not to exceed $50,000. Later he raised the prize value to $100,000, and in 2013 the billionaire offered $1,000,000 for a solution.

The Heegner Numbers

Is there a simple formula that generates only primes? That was the hope of early lovers of primes. Euler came up with a marvelous observation: the polynomial $x^2 - x + 41$ is a prime for all values $x = 1, 2 \ldots, 40$. There are other numbers n besides 41 which have the property that $x^2 - x + n$ is a prime for $x = 1, 2, \ldots, n - 1$: these are the numbers $n = 2, 3, 5, 11$, and 17. Are there other values of n that have this property, or is 41 the largest? This question has an interesting story, which requires moving to an apparently different topic.

We already encountered Fermat's Two-Squares Theorem concerning when a number n could be written in the form $n = x^2 + y^2$. More generally, one wonders about the values produced by the quadratic form $ax^2 + bxy + cy^2$. Now let's do some algebraic magic. If the numbers s, t, u, and v satisfy $sv - tu = 1$, then we construct new numbers x' and y' through the equations

$$\begin{pmatrix} x \\ y \end{pmatrix} = \begin{pmatrix} s & t \\ u & v \end{pmatrix} \begin{pmatrix} x' \\ y' \end{pmatrix}$$

By expanding, one verifies that

$$ax^2 + bxy + cy^2 = A\left(x'\right)^2 + Bx'y' + C\left(y'\right)^2 \tag{9.3}$$

where

$$A = as^2 + bsu + cu^2$$

$$B = 2ast + b(sv + tu) + 2cuv$$

$$C = at^2 + btv + cv^2$$

Equation (9.3) implies that the two quadratic forms with coefficients (a, b, c) and (A, B, C) represent the same set of integers. We say that these two forms are *equivalent*.

A little more algebraic sleight of hand reveals that two equivalent forms have the same discriminant: $D = b^2 - 4ac = B^2 - 4AC < 0$. This all led researchers—both Lagrange and Gauss played major roles here—to define the *class number* of D to be the number of inequivalent quadratic forms $ax^2 + bxy + cy^2$ whose discriminant $b^2 - 4ac$ equals D. Of particular interest was finding how many values of D had a class number 1.

Work on the class number problem progressed incrementally over many years. In the 1930s, it was shown that only a finite number of discriminants have a prescribed class number and that there were at most 10 values of D that had a class number of 1: the nine numbers

$$-3, -4, -7, -8, -11, -19, -43, -67, -163$$

and at most one other value. In the 1950s, high school teacher Kurt Heegner claimed he had proved that the hypothetical 10th number did not exist. His paper had some errors, and unfortunately, he passed away before his work was properly understood. In 1967, Harold Stark used an approach similar to Heegner's to produce a proof that was accepted. It should be mentioned that in the previous year, Alan Baker proved the same result using a completely different approach. In any case, the nine numbers that have a class number of 1 are called the *Heegner numbers.*

How does the class number problem relate to sequences of primes? A 1912 theorem, presented by G. Rabinovich at the Fifth International Congress of Mathematicians, makes the connection. If $D < 0$ and $D \equiv 1 \bmod 4$, then the numbers

$$x^2 - x + \frac{1 + |D|}{4}$$

are prime for $x = 1, 2, \ldots, \frac{|D|-3}{4}$ if and only if the class number of D equals 1. The Heegner number -163 shows that $(1 + 163)/4 = 41$ is the largest number that produces a sequence of primes like the one Euler built.

The Heegner numbers have deep connections to other parts of number theory. Here are some eye-popping approximations:

$$e^{\pi\sqrt{19}} \approx 12^3(3^2 - 1)^3 + 744 - 0.22$$

$$e^{\pi\sqrt{43}} \approx 12^3(9^2 - 1)^3 + 744 - 0.00022$$

$$e^{\pi\sqrt{67}} \approx 12^3(21^2 - 1)^3 + 744 - 0.0000013$$

$$e^{\pi\sqrt{163}} \approx 12^3(231^2 - 1)^3 + 744 - 0.00000000000075$$

The approximation for $e^{\pi\sqrt{163}}$ was already noted by Charles Hermite in 1859. In 1975, Martin Gardner, as an April Fool's joke, claimed that this number was an integer and had been proved by Ramanujan. It has therefore been called *Ramanujan's constant.*

10

Solutions

Rearranging Four Pieces (Chapter 4)

The four pieces cannot be assembled together into the new rectangle of figure 4.4. A clue is that the slope of the diagonal is simultaneously 3/8 and 5/13, clearly a problem!

The Four Hats Problem (Chapter 4)

The answer is that C knows his hat is black. How is this possible? All he sees is that B's hat is white. If his own hat was white, D would see the two white hats and immediately know that his own hat is black. Since he is **not** calling out—C waits one minute to give him a chance—C knows that his own hat must be black.

For the variant of the problem, suppose that one of B, C, or D is wearing the black hat. Since the others on that side see the black hat on someone else, each knows that his own hat is white and either of them would immediately call out. If these three are all wearing white hats, none of them will call out, so after waiting a minute, everyone (including student A) knows that A is wearing the black hat.

The Kuratowski Closure–Complement Theorem (Chapter 7)

Letting $\mathbb{Q}$ denote the rational numbers, the set

$$S = (0,1) \cup (1,2) \cup 3 \cup ([4,5] \cap \mathbb{Q})$$

produces 14 distinct closures and complements:

$$S = (0,1) \cup (1,2) \cup \{3\} \cup ([4,5] \cap \mathbb{Q})$$

$$cS = [0,2] \cup \{3\} \cup [4,5]$$

$$kcS = (-\infty,0) \cup (2,3) \cup (3,4) \cup (5,\infty)$$

$$ckcS = (-\infty,0] \cup [2,4] \cup [5,\infty)$$

$$kckcS = (0,2) \cup (4,5)$$

$$ckckcS = [0,2] \cup [4,5]$$

$$kckckcS = (-\infty,0) \cup (2,4) \cup (5,\infty)$$

$$kS = (-\infty,0) \cup \{1\} \cup [2,3) \cup (3,4) \cup ((4,5) \cap k\mathbb{Q}) \cup (5,\infty)$$

$$ckS = (\infty,0] \cup \{1\} \cup [2,\infty)$$

$$kckS = (0,1) \cup (1,2)$$

$$ckckS = [0,2]$$

$$kckckS = (\infty,0) \cup (2,\infty)$$

$$ckckckS = (\infty,0] \cup [2,\infty)$$

$$kckckckS = (0,2)$$

Recreational Mathematics (Chapter 7)

If the digits of a number n are raised to the kth power and their sum is n itself, we call n a *k-narcissistic number*. Thus, the number $14,459,929$ is 7-narcissistic. There are four 3-narcissistic numbers: 153, 370, 371, and 407.

further reading

Many of the topics in this book are well established and can be found in mathematics textbooks, while others require digging into research journals. An online search is the best way to get started to find more details about a given topic. This is a list of accessible books that allow one to further explore at least one of the topics covered in the current volume.

Jörg Arndt and Christoph Haenel, *Pi Unleashed*, Springer-Verlag, New York, 2000.

Emil Artin, *The Gamma Function*, Holt, Rinehart and Winston, New York, 1964.

John Baez, "The Octonions," *Bull. Amer. Math. Soc.* 39: 145–205.

E. R. Berlekamp, J. H. Conway, and R. K. Guy, *Winning Ways for Your Mathematical Plays*, A. K. Peters, CRC Press, Boca Raton, FL, 2001–2004.

B. Bollobas (editor), *Littlewood's Miscellany*, Cambridge University Press, Cambridge, U.K., 1990.

Jonathan Borwein and David Bailey, *Mathematics by Experiment*, A. K. Peters, CRC Press, Boca Raton, FL, 2004.

J. H. Conway and R. K. Guy, *The Book of Numbers*, Springer, New York, 1996.

H.S.M. Coxeter and S. L. Greitzer, *Geometry Revisited*, Mathematical Association of America, Washington, DC, 1967.

Joseph W. Dauben, *Georg Cantor: His Mathematics and Philosophy of the Infinite*, Harvard University Press, Cambridge, MA, 1979.

Joseph W. Dauben, "Georg Cantor and the Battle for Transfinite Set Theory," *Proceedings of the 9th ACMS Conference* (Westmont College, Santa Barbara, CA), 1993 and 2005, pp. 1–22.

Joseph W. Dauben, "Georg Cantor and Pope Leo XIII: Mathematics, Theology, and the Infinite," *Journal of the History of Ideas* 38 (1): (1977), pp. 85–108.

Philip J. Davis, Reuben Hersh, Elena Anne Marchisotto, *The Mathematical Experience*, Birkhäuser, Boston, 1995.

Erik D. Demaine and J. O'Rourke, *Geometric Folding Algorithms: Linkages, Origami, Polyhedra*, Cambridge University Press, New York, 2007.

Apostolos Doxiadis, *Uncle Petros and Goldbach's Conjecture: A Novel of Mathematical Obsession*, Bloomsbury, New York, 2001.

Underwood Dudley, *The Trisectors*, Mathematical Association of America, Washington, DC, 1996.

Martin Gardner, "The Fantastic Combinations of John Conway's New Solitaire Game 'Life.'" *Scientific American* 223 (October 1970), pp. 120–123.

Martin Gardner, *Hexaflexagons and Other Mathematical Diversions*, Simon and Schuster, New York, 1959.

G. H. Hardy, *Ramanujan*, Cambridge University Press, Cambridge, UK, 1940.

G. H. Hardy and E.M. Wright, *An Introduction to the Theory of Numbers* (6th edition), Oxford University Press, New York, 2008.

Julian Havil, *The Irrationals*, Princeton University Press, Princeton, NJ, 2012.

Fukagawa Hidetoshi and Tony Rothman, *Sacred Mathematics: Japanese Temple Geometry*, Princeton University Press, Princeton, NJ, 2008.

David Hilbert, "Über das Unendliche." *Mathematische Annalen* 95 (1926): 161–190.

Paul Hoffman, *The Man Who Loved Only Numbers*, Hyperion, New York, 1999.

D. A. Holton and J. Sheehan, *The Petersen Graph*, Australian Mathematical Society Lecture Series (Book 7), Cambridge University Press, New York, 1993.

Dan Kalman, "The Most Marvelous Theorem in Mathematics," *The Journal of Online Mathematics and Its Applications*, Volume 8 (March 2008).

Robert Kanigel, *The Man Who Knew Infinity: A Life of the Genius Ramanujan*, Washington Square Press, New York, 1992.

Victor Klee and Stan Wagon, *Old and New Unsolved Problems in Plane Geometry and Number Theory*, The Dolciani Mathematical Expositions, No. 11, Mathematical Association of America, Washington, DC, 1991.

Jeffrey C. Lagarias (editor), *The Ultimate Challenge: The 3x+1 Problem*, American Mathematical Society, Providence, RI, 2010.

"Landau's Problems," *Wikipedia*, last modified on 10 November 2014 at 23:17, http://en.wikipedia.org/wiki/Landau%27s_problems

Harold W. Lewis, *Why Flip a Coin?* John Wiley, Hoboken, NJ, 1997.

Dana Mackenzie, "The Poincaré Conjecture—Proved," *Science* 22: 314, no. 5807, (December 2006), pp. 1848–1849.

Stanley Milgram, "The Small World Problem," *Psychology Today* 1(1), May 1967, pp. 61–67.

J. O'Rourke, *Art Gallery Theorems and Algorithms*, Oxford University Press, New York, 1987.

Donal O'Shea, *The Poincaré Conjecture*, Walker & Company, New York, 2007.

Heinz-Otto Peitgen, Hartmut Jürgens, and Dietmar Saupe, *Chaos and Fractals*, 2nd edition, Springer, 2004.

Ivars Peterson, "The Honeycomb Conjecture," *Science News*, 156 (4), (July 24, 1999), p. 60–61.

David S. Richeson, *Euler's Gem*, Princeton University Press, Princeton, NJ, 2012.

George Finlay Simmons, *Calculus Gems*, McGraw Hill, New York, 1992.

Simon Singh, *Fermat's Enigma*, Anchor, New York, 1998.

Statement of Philosophy and Criteria for the Journal Experimental Mathematics. *Journal Experimental Mathematics.* http://www.emis.de/journals/EM/expmath/philosophy.html

George G. Szpiro, *Kepler's Conjecture*, Wiley, Hoboken, NJ, 2003.

G. Szpiro, "Does the Proof Stack Up?" *Nature* 424, 2003, pp. 12–13.

Daniel Tammet, *Born on a Blue Day*, Free Press, New York, 2006.

B. Thwaites, "Two Conjectures, or How to Win £1100." *Math. Gaz.* 80, (1996), pp. 35–36.

Robin Wilson, *Four Colors Suffice*, Princeton University Press, Princeton, NJ, 2004.

T. Y. Yi and J. A. Yorke, "Period Three Implies Chaos." *Amer. Math. Monthly* 82 (1975): 985–992.

credits for illustrations

FIGURE 1.7. From Eli Maor and Eugen Jost. *Beautiful Geometry.* © 2014 Princeton University Press. Reproduced with permission.

FIGURE 1.9. Bernsley fern courtesy of W. Garrett Mitchener, College of Charleston Mathematics Department. Menger sponge from Timothy P. Chartier, *Math Bytes: Google Bombs, Chocolate-Covered Pi, and Other Cool Bits in Computing.* © 2014 Princeton University Press. Reproduced with permission.

FIGURE 2.1. Mona Lisa image courtesy of Bob Bosch.

FIGURE 2.9. Tower of Hanoi, solution with four disks: courtesy of Eric W. Weisstein, "Tower of Hanoi." From MathWorld—A Wolfram Web Resource.

FIGURE 2.11. Courtesy of Mathematikum Giessen.

FIGURE 2.12. From Satyan L. Devadoss and Joseph O'Rourke, *Discrete and Computational Geometry.* © 2011 Princeton University Press. Reproduced with permission.

FIGURE 2.15. Courtesy of Todd Stedl.

FIGURE 3.14. Courtesy of Josh Pesavento

FIGURE 3.18. Three-dimensional proof courtesy of National Tsing Hua University, Taiwan.

FIGURE 4.5. Both images courtesy of Archives of the Mathematisches Forschungsinstitut Oberwolfach.

Figure 4.6. Courtesy of Yassine Mrabet

Figure 4.7. Courtesy of Eric W. Weisstein, "Villarceau Circles." From MathWorld—A Wolfram Web Resource. http://mathworld.wolfram.com/VillarceauCircles.html

Figure 5.7. Knight's tour on a 24x24 chessboard generated by a neural network algorithm, by "Pattern86." Licensed under the Creative Commons Attribution-Share Alike 3.0 Unported license (http://creativecommons.org/licenses/by-sa/3.0/deed.en)

Figure 5.8. Geodesic dome courtesy of Philipp Hienstorfer.

Figure 5.9. From John J. Watkins, *Number Theory: A Historical Approach*. © 2014 Princeton University Press. Reproduced with permission.

Figure 6.1. "Honeycombs made of wax and full of honey, built by black bees" (Apis mellifera mellifera). Courtesy of Emmanuel Boutet.

Figure 6.2. Oranges in the Borough Market, London. Courtesy of José Luis Sánchez Mesa.

Figure 6.4. Courtesy of Eric W. Weisstein, "Hexlet." From MathWorld— A Wolfram Web Resource. http://mathworld.wolfram.com/Hexlet.html

Figure 6.6. Hex game © Dvortygirl. Licensed under Creative Commons Attribution-ShareAlike 2.0 Generic (https://creativecommons.org/licenses/by-sa/2.0/)

Figure 7.7. Quadratic Chladni plate (http://commons.wikimedia.org/wiki/File:Quadratic_Chladni_plate.JPG).

Figure 8.1. Proof of Pizza Theorem. Proof without words of the pizza theorem based on Larry Carter and Stan Wagon, (1994a), "Proof without Words: Fair Allocation of a Pizza," *Mathematics Magazine* 67 (4): 267, adapted by "dmcq" and licensed under the Creative Commons Attribution-Share Alike 3.0 Unported license (http://creativecommons.org/licenses/by-sa/3.0/deed.en)

Figure 8.5. Courtesy of Clinton Curry.

Figure 8.7. Courtesy of Claudio Rocchinin. Licensed under CC-BY-2.5.

Figure 9.2. Courtesy of Toby Hudson.

index